I0038412

# Food Processing: Principles, Technologies and Applications

# Food Processing: Principles, Technologies and Applications

Zachary Walker

Larsen & Keller
www.larsen-keller.com

Food Processing: Principles, Technologies and Applications
Zachary Walker
ISBN: 978-1-64172-487-6 (Hardback)

© 2020 Larsen & Keller

# ▤ Larsen & Keller

Published by Larsen and Keller Education,
5 Penn Plaza,
19th Floor,
New York, NY 10001, USA

**Cataloging-in-Publication Data**

Food processing : principles, technologies and applications / Zachary Walker.
    p. cm.
Includes bibliographical references and index.
ISBN 978-1-64172-487-6
1. Food industry and trade. 2. Food industry and trade--Technological innovations.
3. Food processing plants. 4. Processed foods. I. Walker, Zachary.
TP370.5 .F66 2020
664--dc23

This book contains information obtained from authentic and highly regarded sources. All chapters are published with permission under the Creative Commons Attribution Share Alike License or equivalent. A wide variety of references are listed. Permissions and sources are indicated; for detailed attributions, please refer to the permissions page. Reasonable efforts have been made to publish reliable data and information, but the authors, editors and publisher cannot assume any responsibility for the validity of all materials or the consequences of their use.

Trademark Notice: All trademarks used herein are the property of their respective owners. The use of any trademark in this text does not vest in the author or publisher any trademark ownership rights in such trademarks, nor does the use of such trademarks imply any affiliation with or endorsement of this book by such owners.

For more information regarding Larsen and Keller Education and its products, please visit the publisher's website www.larsen-keller.com

# Table of Contents

**Permissions**

**Index**

# Preface

Food processing is the set of techniques and operations, which are used in the transformation of raw ingredients into food that are suitable for consumption. It includes complex industrial methods for making convenience foods. Food processing is classified into primary, secondary and tertiary food processing. Primary food processing converts agricultural products into a product, which can be consumed. Secondary food processing produces food which are ready to use, for example, baked bread. Commercial production of processed food is known as tertiary food processing. Common techniques of food processing include fermentation, liquefaction, pasteurization, food irradiation, etc. This book outlines the processes and applications of food processing in detail. The topics included herein on food processing are of utmost significance and bound to provide incredible insights to readers. Those in search of information to further their knowledge will be greatly assisted by this book.

A detailed account of the significant topics covered in this book is provided below:

Chapter 1- Food processing is defined as the process of transforming raw ingredients into edible foods for consumption of humans and animals. Its methods include chopping, slicing, liquefaction, fermentation, emulsification, mixing, pasteurization, packaging, etc. This is an introductory chapter which will briefly introduce all these methods of food processing as well as their hygienic designs.

Chapter 2- Ambient temperature processing is the method of food processing that destroys the microbial growth without any change in the temperature of the surrounding environment. Centrifugation, size reduction, coating of food and mixing of food are some of the concepts that fall within this processing. This chapter has been carefully written to provide an easy understanding of these concepts of ambient temperature processing.

Chapter 3- Heat processing is defined as the food processing technique that makes use of high temperatures to sterilize food. It includes blanching, pasteurization, evaporation, distillation, dehydration, smoking, frying, extrusion cooking, etc. All these methods related to heat processing have been carefully analyzed in this chapter.

Chapter 4- The preservation of food by lowering the temperature helps in maintaining the nutritional value and sensory characteristics of food. Some of the processes used are refrigeration, cooling, freezing, freeze-drying, etc. This chapter closely examines these processes to provide an extensive understanding of the subject.

Chapter 5- Post-processing operations are used in improving the quality of food and extend shelf life of some processed foods. Different operations such as materials handling, packaging, sealing, storage, etc. are used in the post processing of food. The topics elaborated in this chapter will help in gaining a better perspective about these post-processing operations.

I would like to make a special mention of my publisher who considered me worthy of this opportunity and also supported me throughout the process. I would also like to thank the editing team at the back-end who extended their help whenever required.

**Zachary Walker**

# Food Processing: An Introduction

Food processing is defined as the process of transforming raw ingredients into edible foods for consumption of humans and animals. Its methods include chopping, slicing, liquefaction, fermentation, emulsification, mixing, pasteurization, packaging, etc. This is an introductory chapter which will briefly introduce all these methods of food processing as well as their hygienic designs.

## Food Processing

Food processing is the set of methods and techniques used to transform raw ingredients into food or food into other forms for consumption by humans or animals either in the home or by the food processing industry. Food processing typically takes clean, harvested crops or slaughtered and butchered animal products and uses these to produce attractive, marketable, and often long-life food products. Similar processes are used to produce animal feed. Extreme examples of food processing include the expert removal of toxic portions of the fugu fish or preparing space food for consumption under zero gravity.

Examples of some processed foods.

The benefits of food processing include the preservation, distribution, and marketing of food, protection from pathogenic microbes and toxic substances, year-round availability of many food items, and ease of preparation by the consumer. On the other hand, food processing can lower the nutritional value of foods, and processed foods may include additives (such as colorings, flavorings, and preservatives) that may have adverse health effects.

## Food Processing Methods

Common food processing techniques include:

- Removal of unwanted outer layers, such as potato peeling or the skinning of peaches;

- Chopping or slicing, such as to produce diced carrots;

- Mincing and macerating;

- Liquefaction, such as to produce fruit juice;

- Fermentation, as in beer breweries;

- Emulsification;

- Cooking, by methods such as baking, boiling, broiling, frying, steaming, or grilling;

- Mixing;

- Addition of gas such as air entrainment for bread or gasification of soft drinks;

- Proofing;

- Spray drying;

- Pasteurization;

- Packaging.

## Performance Parameters for Food Processing

When designing processes for the food industry, the following performance parameters may be taken into account:

- Hygiene, measured, for instance, by the number of microorganisms per ml of finished product.

- Energy consumption, measured, for instance, by "ton of steam per ton of sugar produced".

- Minimization of waste, measured, for instance, by the "percentage of peeling loss during the peeling of potatoes".

- Labor used, measured, for instance, by the "number of working hours per ton of finished product".

- Minimization of cleaning stops, measured, for instance, by the "number of hours between cleaning stops".

# Benefits

More and more people live in the cities far away from where food is grown and pro-duced. In many families, the adults are work from home and therefore there is little time for the preparation of food based on fresh ingredients. The food industry offers products that fulfill many different needs: from peeled potatoes that simply need to be boiled at home to fully prepared ready meals that can be heated up in the microwave oven in a few minutes.

Benefits of food processing include toxin removal, preservation, easing marketing and distribution tasks, and increasing food consistency. In addition, it increases seasonal availability of many foods, enables transportation of delicate perishable foods across long distances, and makes many kinds of foods safe to eat by de-activating spoilage and pathogenic micro-organisms. Modern supermarkets would not be feasible without modern food processing techniques, long voyages would not be possible, and military campaigns would be significantly more difficult and costly to execute.

Modern food processing also improves the quality of life for allergy sufferers, diabetics, and other people who cannot consume some common food elements. Food processing can also add extra nutrients such as vitamins.

Processed foods are often less susceptible to early spoilage than fresh foods, and are better suited for long distance transportation from the source to the consumer. Fresh materials, such as fresh produce and raw meats, are more likely to harbor pathogenic microorganisms (for example, Salmonella) capable of causing serious illnesses.

# Drawbacks

In general, fresh food that has not been processed other than by washing and sim-ple kitchen preparation, may be expected to contain a higher proportion of natu-rally occurring vitamins, fiber and minerals than the equivalent product processed by the food industry. Vitamin C, for example, is destroyed by heat and therefore canned fruits have a lower content of vitamin C than fresh ones.

Food processing can lower the nutritional value of foods. Processed foods tend to in-clude food additives, such as flavorings and texture enhancing agents, which may have little or no nutritive value, and some may be unhealthy. Some preservatives added or created during processing, such as nitrites or sulfites, may cause adverse health effects.

Processed foods often have a higher ratio of calories to other essential nutrients than unprocessed foods, a phenomenon referred to as "empty calories." Most junk foods are processed, and fit this category.

High quality and hygiene standards must be maintained to ensure consumer safety, and failure to maintain adequate standards can have serious health consequences.

Processing food is a very costly process, thus increasing the prices of foods products.

## Trends in Modern Food Processing

## Health

- Reduction of fat content in final product, for example, by using baking instead of deep-frying in the production of potato chips.

- Maintaining the natural taste of the product, for example, by using less artificial sweetener.

## Hygiene

The rigorous application of industry and government endorsed standards to minimize possible risk and hazards. In the U.S., the standard adopted is HACCP.

## Efficiency

- Rising energy costs lead to increasing usage of energy-saving technologies, for example, frequency converters on electrical drives, heat insulation of factory buildings, and heated vessels, energy recovery systems.

- Factory automation systems (often Distributed control systems) reduce personnel costs and may lead to more stable production results.

## Industries

Food processing industries and practices include the following:

- Cannery;
- Industrial rendering;
- Meat packing plant;
- Slaughterhouse;
- Sugar industry;
- Vegetable packing plant.

# Basic Principles of Processing

Processing of foods is a segment of manufacturing industry that transforms animal, plant, and marine materials into intermediate or finished value-added food products that are safer to eat. This requires the application of labor, energy, machinery, and scientific knowledge to a step (unit operation) or a series of steps (process) in achieving the desired transformation. Value-added ingredients or finished products that satisfy consumer needs and convenience are obtained from the raw materials.

The aims of food processing could be considered four-fold: extending the period during which food remains wholesome (microbial and biochemical), providing (supplementing) nutrients required for health, providing variety and convenience in diet, and adding value.

Food materials' shelf life extension is achieved by preserving the product against biological, chemical, and physical hazards. Bacteria, viruses, and parasites are the three major groups of biological hazards that may pose a risk in processed foods. Biological hazards that may be present in the raw food material include both pathogenic microorganisms with public health implications and spoilage microorganisms with quality and esthetic implications. Mycotoxin, pesticide, fungicide, and allergens are some examples of chemical hazards that may be present in food. Physical hazards may involve the presence of extraneous material (such as stones, dirt, metal, glass, insect fragments, hair). These hazards may accidentally or deliberately (in cases of adulteration) become part of the processed product. Food processing operations ensure targeted removal of these hazards so that consumers enjoy safe, nutritious, wholesome foods. With the possibility of extending shelf life of foods and advances in packaging technology, food processing has been catering to consumer convenience by creating products, for example, ready-to-eat breakfast foods and TV dinners, on-the-go beverages and snacks, pet foods, etc. Food processing, as an industry, has also responded to changes in demographics by bringing out ethnic and specialty foods and foods for elderly people and babies. Nutrition fortification, for example, folic acid supplementation in wheat flour, is another function of processing food.

The scope of food processing is broad; unit operations occurring after harvest of raw materials until they are processed into food products, packaged, and shipped for retailing could be considered part of food processing. Typical processing operations may include raw material handling, ingredient formulation, heating and cooling, cooking, freezing, shaping, and packaging. These could broadly be categorized into primary and secondary processing. Primary processing is the processing of food that occurs after harvesting or slaughter to make food ready for consumption or use in other food products. Primary processing ensures that foods are easily transported and are ready to be sold, eaten or processed into other products (e.g. after the primary processing of peeling and slicing, an apple can be eaten fresh or baked into a pie). Secondary processing turns the primary-processed food or ingredient into other food products. It ensures that foods can be used for a number of purposes, do not spoil quickly, are healthy and wholesome to eat, and are available all year (e.g. seasonal foods).

In the previous example, baking of the pie is a secondary processing step, which utilizes ingredient from primary processing (sliced apple).

The food and beverage manufacturing industry is one of the largest manufacturing sectors in the US. In 2011, these plants accounted for 14.7% of the value of shipments from all US manufacturing plants. Meat processing is the largest single component of food and beverage manufacturing, with 24% of shipments in 2011. Other important components include dairy (13%), beverages (12%), grains and oilseeds (12%), fruits and vegetables (8%), and other food products (11%). Meat processing is also the largest component (17%) of the food sector's total value added, followed by beverage manufacturing (16%). California has the largest number of food manufacturing plants, followed by New York and Texas. Demand for processed foods tends to be less susceptible to fluctuating economic conditions than other industries.

Some basic principles associated with processing and preservation of food are summarized here.

## Unit Operations in Food Processing

Most food processes utilize six different unit operations: heat transfer, fluid flow, mass transfer, mixing, size adjustment (reduction or enlargement), and separation.

During food processing, food material may be combined with a variety of ingredients (sugar, preservatives, acidity) to formulate the product and then subjected to different unit operations either sequentially or simultaneously. Food processors often use process flow charts to visualize the sequence of operations needed to transform raw materials into final processed product. The process flow diagrams often include quality control limits and/or adjustment and description of any hazards. Figure shows a sample process flow diagram for making Frankfurter comminuted sausage.

### Heat Transfer

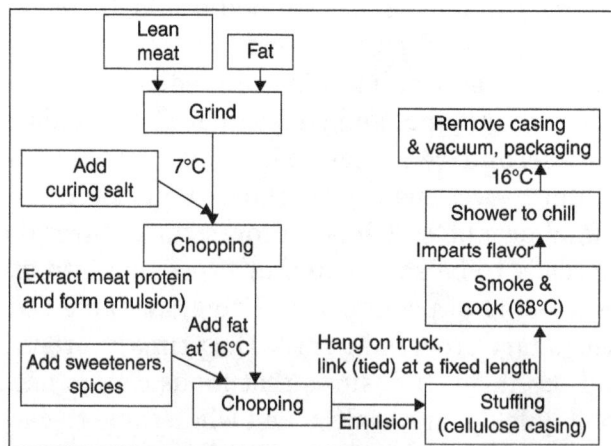

Process flow diagram of Frankfurter comminuted sausage manufacturing.

Heat transfer is one of the fundamental processing principles applied in the food industry and has applications in various unit operations, thermal processing, evaporation (concentration) and drying, freezing and thawing, baking, and cooking. Heating is used to destroy microorganisms to provide a healthy food, prolong shelf life through the destruction of certain enzymes, and promote a product with acceptable taste, odor, and appearance. Heat transfer is governed by heat exchange between a product and its surrounding medium. The extent of heat transfer generally increases with increasing temperature difference between the product and its surrounding.

Conduction, convection, and radiation are the three basic modes of heat transfer. Conduction heat transfer occurs within solid foods, wherein a transfer of energy occurs from one molecule to another. Generally, heat energy is exchanged from molecules with greater thermal energy to molecules located in cooler regions. Heat transfer within a potato slice is an example of conduction heat transfer.

Heat is transferred in fluid foods by bulk movement of fluids as a result of a temperature gradient, and this process is referred to as convective heat transfer. Convective heat transfer can be further classified as natural convection and forced convection. Natural convection is a physical phenomenon wherein a thermal gradient due to density difference in a heated product causes bulk fluid movement and heat transfer. Movement of liquids inside canned foods during thermal sterilization is an example of natural convection. If the movement and heat transfer are facilitated by mechanical agitation (such as use of mixers), this is called forced convection.

Radiation heat transfer occurs between two surfaces as a result of the transfer of heat energy by electromagnetic waves. This mode of heat transfer does not require a physical medium and can occur in a vacuum. Baking is one example of heat transfer via radiation from the heat source in the oven to the surface of bread. However, heat propagates via conduction within the body of the bread.

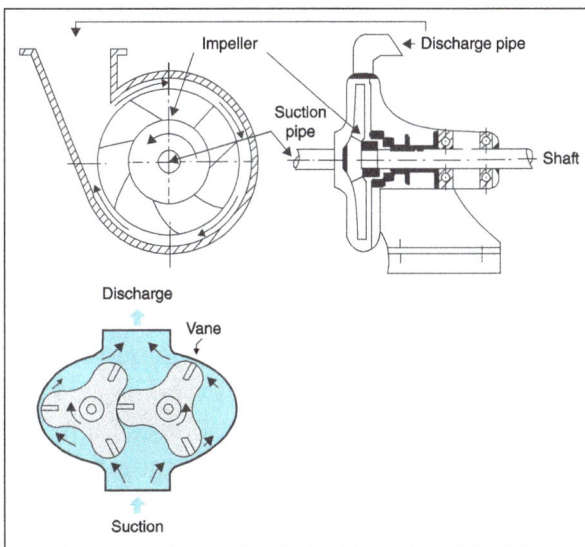

Schematic of centrifugal.

## Mass Transfer

Mass transfer involves migration of a constituent of fluid or a component of a mixture in or out of a food product. Mass transfer is controlled by the diffusion of the component within the mixture. The mass migration occurs due to changes in physical equilibrium of the system caused by concentration or vapor pressure differences. The mass transfer may occur within one phase or may involve transfer from one phase to another. Food process unit operations that utilize mass transfer include distillation, gas absorption, crystallization, membrane processes, evaporation, and drying.

## Fluid Flow

Fluid flow involves transporting liquid food through pipes during processing. Powders and small-particulate foods are handled by pneumatic conveying, whereas fluids are transported by gravity flow or through the use of pumps. The centrifugal pump and the positive displacement pump are two pumps commonly used for fluid flow.

Centrifugal pumps utilize a rotating impeller to create a centrifugal force within the pump cavity, so that the fluid is accelerated until it attains its tangential velocity close to the impeller tip. The flow is controlled by the choice of impeller diameter and rotary speed of the pump drive. The product viscosity is an important factor affecting centrifugal pump performance; if the product is sufficiently viscous, the pump cavity will not fill with every revolution and the efficiency of the pump will be greatly reduced. Centrifugal pumps are used for transportation of fluids from point A to point B as in transporting fluid for cleaning operations. Centrifugal pumps do not have a constant flow rate.

A positive displacement pump generally consists of a reciprocating or rotating cavity between two lobes or gears and a rotor. Fluid enters by gravity or a difference in pressure and the fluid forms the seals between the rotating parts. The rotating movement of the rotor produces the pressure to cause the fluid to flow. Because there is no frictional loss, positive pumps are used where a constant rate of flow is required (timing pump), for high-viscosity fluids or for transporting fragile solids suspended in a fluid (such as moving cottage cheese curd from a vat to a filler).

## Mixing

Mixing is a common unit operation used to evenly distribute each ingredient during manufacturing of a food product. Mixing is generally required to achieve uniformity in the raw material or intermediate product before it is taken for final production. Mixing of cookie or bread dough is an example, wherein required ingredients need to be mixed well into a uniform dough before they are portioned into individual cookies or loaves. Application of mechanical force to move ingredients (agitation) generally accomplishes this goal. Efficient heat transfer and/or uniform ingredient incorporation are two goals of mixing. Different mixer configurations can be used to achieve different purposes.

The efficiency of mixing depends upon the design of impeller, including its diameter, and the speed baffle configurations.

## Size Adjustment

In size adjustment, the food is reduced mostly into smaller pieces during processing, as the raw material may not be at a desired size. This may involve slicing, dicing, cutting, grinding, etc. However, increasing a product size is also possible. For example, aggregation, agglomeration (instant coffee), and gelation are examples of size adjustment that results in increase in size. In the case of liquid foods, size reduction is often achieved by homogenization. During milk processing, fats are broken into emulsions via homogenization for further separation.

## Separation

This aspect of food processing involves separation and recovery of targeted food components from a complex mixture of compounds. This may involve separating a solid from a solid (e.g. peeling of potatoes or shelling of nuts), separating a solid from a liquid (e.g. filtration, extraction) or separating liquid from liquid (e.g. evaporation, distillation). Industrial examples of separation include crystallization and distillation, sieving, and osmotic concentration. Separation is often used as an intermediate processing step, and is not intended to preserve the food.

## Thermophysical Properties, Microbial Aspects and other Considerations in Food Processing

### Raw Material Handling

Raw material handling is the very first step in the food processing. Raw material handling includes postharvest transportation (farm to plant), sorting, cleaning or sanitizing before loading into equipment in the plant. These could also be considered as part of primary processing of the food materials. Microorganisms could attach to inert non-porous surfaces in raw foods and it has been demonstrated that these cells transfer from one surface to another to another when contact occurs. Appropriate raw material selection and handling affect microbial safety and final product quality. Future food preservation studies need to consider the impact of raw material (including postharvest handling prior to preservation) on the final processed product.

### Cleaning and Sanitation

Cleaning and sanitation of raw food material could be considered the first step in controlling any contamination of foreign materials or microorganisms during food processing. Cleaning removes foreign materials (i.e. soil, dirt, animal contaminants) and prevents the accumulation of biological residues that may support the growth of harmful microbes, leading to disease and/or the production of toxins. Sanitization is the use of any chemical or other effective method to reduce the initial bacterial load on the surface of raw materials or food processing equipment. Efficient sanitization includes both the outside and the inside of the plant such as specific floor plan, approved materials used in construction, adequate light, air ventilation, direction of air flow, separation of processing areas for raw and finished products, sufficient space for operation and movement, approved plumbing, water supply, sewage disposal system, waste treatment facilities, drainage, soil conditions, and the surrounding environment.

## Engineering Properties of Food, Biological and Packaging Material

Knowledge of various engineering (physical, thermal, and thermodynamic) properties of food, biological, and packaging material is critical for successful product development, quality control, and optimization of food processing operations. For example, data on density of food material are important for separation, size reduction or mixing processes. Knowledge of thermal properties of food (thermal conductivity, specific heat, thermal diffusivity) is useful in identifying the extent of process uniformity during thermal processes such as pasteurization and sterilization. For liquid foods, knowledge of rheological characteristics, including viscosity, helps in the design of pumping systems for different continuous flow operations. Different food process operations (heating, cooling, concentration) can alter product viscosity during processing, and this needs to be considered during design. Phase and glass transition characteristics of food materials govern many food processing operations such as freezing, dehydration, evaporation, and distillation. For example, the density of water decreases when the food material is frozen and as a result increases product volume. This should be considered when designing freezing operations. Thus, food scientists and process engineers need to adequately characterize or gather information about relevant thermophysical properties of food materials being processed.

## Microbiological Considerations

Most raw food materials naturally contain microorganisms, which bring both desirable and undesirable effects to processed food. For example, many fermented foods (e.g. ripened cheeses, pickles, sauerkraut, and fermented sausages) have considerably extended shelf life, developed aroma, andflavor characteristics over those of the raw materials arising from microorganisms such as Lactobacillus, Lactococcus, and Staphylococcus bacteria. On the other hand, raw food material also contains pathogens and spoilage organisms. Different foods harbor different pathogens and spoilage organisms. For example, raw apple juice or cider may be contaminated with Escherichia coli O157:H7. Listeria monocytogenes are pathogens of concern in milk and ready-to-eat meat. The target pathogen of concern in shelf-stable low-acid foods (such as soups) is Clostridium botulinum spores. Different pathogenic and spoilage microorganisms offer varied degrees of resistance to thermal treatment. Accordingly, the design of an adequate process to produce safer products depends in part on the resistance of such microorganisms to lethal agents, food material, and desired shelf life.

# Role of Acidity and Water Activity in Food Safety and Quality

Intrinsic food properties (e.g. water activity, acidity, redox potential) can play a role in determining the extent of food processing operations needed to ensure food safety and minimize quality abuse.

Higher acidity levels (pH <4.6) are often detrimental to the survival of microorganisms, so milder treatments are sufficient to preserve an acidic food. Low-acid foods (pH ≥4.6) support the growth and toxin production of various pathogenic microorganisms, including Clostridium botulinum. Products such as milk, meat, vegetables, and soups are examples of low-acid foods and require more severe heat treatment than acid foods such as orange juice or tomato products. pH of the food material also impacts many food quality attributes such as color, texture, and flavor. For example, pH of the milk used for cheese manufacturing can help determine cheese texture (hard/soft). Similarly, pH of fruit jelly can determine gel consistency.

Knowledge of availability of water for microbial, enzymatic or chemical activity helps predict the stability and shelf life of processed foods. This is often reported as water activity ($a_w$), and is defined as the ratio between partial pressure of water vapor ($p_w$) of the food and the vapor pressure of saturated water ($p_w'$) at the same temperature. The water activity concept is often used in food processing to predict growth of bacteria, yeast, and molds. Bacteria grow mostly between aw values of 0.9 and 1, most enzymes and fungi have a lower aw limit of 0.8, and for most yeasts 0.6 is the lower limit. Thus, food can be made safe by lowering the water activity to a point that will not allow the growth of dangerous pathogens. Foods are generally considered safer against microbial growth at $a_w$ below 0.6. Salt and sugar are commonly used to lower water activity by binding product moisture. In the recent years, there has been increased emphasis on reducing salt in processed foods. However, such changes should be systematically evaluated as salt reduction could potentially compromise microbiological safety and quality of the processed product. Water activity of food material can also influence various chemical reactions. For example, non-enzymatic browning reactions increase with water activity level of 0.6–0.7aw. Similarly, lipid oxidation can be minimized at about water activity level 0.2–0.3.

Table: Decimal reduction time (D value, min) of selected pathogenic and spoilage microorganisms found in food material.

| Pathogenic or spoilage microorganisms | D-value, min* | Temperature, °C | Food |
|---|---|---|---|
| **Bacterial spores** | | | |
| Bacillus stearothermilophilus | 2.1–3.4 | 121 | Phosphate buffer (pH 7.0) |
| Clostridium botulinum (types A and B) | 0.1–0.2 | 121 | |
| Clostridium butyricum | 1.1 | 90 | |

| Clostridium nigrificans | 2–3 | 121 | |
| Bacillus cereus | 1.8–19.1 | 95 | Milk |
| Bacillus coagulans spores | 3.5 | 70–95 | |
| Clostridium sporogenes | 0.1–0.15 | 121 | |
| Clostridium thermosaccharolyticum | 3–4 | 121 | |
| Clostridium perfringens | 6.6 | 100.4 | Beef gravy (pH 7) |
| **Molds** | | | |
| Byssochlamys nivea ascopores | 193 | 80–90 | |
| Neosartorya fischeri ascopores | 15.1 | 85–93 | |
| Talaromyces flavus ascopores | 54 | 70–95 | |
| **Vegetative bacteria** | | | |
| Campylobacter jejuni | 0.74–1 | 55 | Skim milk |
| Escherichia coli O157:H7 | 4.5–6.4 | 57.2 | Ground beef |
| Listeria monocytogenes | 6.27–8.32 | 60 | Beef homogenate |
| Salmonella typhimurium | 396 | 71 | Milk chocolate |
| Vibrio parahaemolyticus | 0.02–0.29 | 55 | Clam homogenate |

*D-value is the time taken to reduce the microbial population by one log-cycle (by 90%) at a given temperature.

## Reaction Kinetics

During processing, the constituents of food undergo a variety of chemical, biological, physical, and sensory changes. Food scientists and engineers need to understand the rate of these changes caused by applying a given processing agent and the resulting modifications, so that they can control process operations to produce a product with the desired quality. Enzyme hydrolysis, browning, and color degradation are examples of chemical changes while inactivation of microorganisms after heat treatment is an example of a biological change. Food engineers rely on microbial and chemical kinetic equations to predict and control various changes happening in the processed food.

Table: Water activity values in different food products.

| Food product | Water activity, $a_w$ |
|---|---|
| Fresh meat and fish | 0.99 |
| Bread | 0.95 |
| Cured ham, medium aged cheese | 0.9 |
| Jams and jellies | 0.80 |
| Plum pudding, fruit cakes, sweetened condensed milk, fruit syrups | 0.80 |
| Rolled oats, fudge, molasses, nuts, fondants | 0.65 |

| Dried fruit | 0.60 |
|---|---|
| Dried foods, spices, noodles | 0.50 |
| Marshmallow | 0.6–0.65 |
| Biscuits | 0.30 |
| Whole milk powder, dried vegetables, corn flakes | 0.20 |
| Instant coffee | 0.20 |
| English toffee | 0.2–0.3 |
| Hard candy | 0.1–0.2 |

## Common Food Preservation/Processing Technologies

## Goals of Food Processing

The food industry utilizes a variety of technologies such as thermal processing, dehydration, refrigeration, and freezing to preserve food materials. The goals of these food preservation methods include eliminating harmful pathogens present in the food and minimizing or eliminating spoilage microorganisms and enzymes for shelf life extension.

The general concepts associated with processing of foods to achieve shelf life extension and preserve quality include addition of heat, removal of heat, removal of moisture, and packaging of foods to maintain the desirable aspects established through processing. Many food processing operations add heat energy to achieve elevated temperatures detrimental to the growth of pathogenic microorganisms. Exposure of food to elevated temperatures for a predetermined length of time (based on the objectives of the process at hand) is a key concept in food processing. Pasteurization of milk, fruit and vegetable juices, canning of plant and animal food products are some examples of processing with heat addition. The microbial inactivation achieved is based on exposure of foods to specific time temperature combinations. Blanching is another example of heat addition, which helps with enzyme inactivation.

Processing of foods by heat removal is aimed more towards achieving shelf life extension by slowing down the biochemical and enzymatic reactions that degrade foods. Removal of moisture is another major processing concept, in which preservation is achieved by reducing free moisture in food to limit or eliminate the growth of spoilage microorganisms. Drying of solid foods and concentration of liquid foods fall under this category. Finally, packaging maintains the product characteristics established by processing of the food, including preventing post processing contamination. Packaging operations are also considered part of food processing.

In recent decades, the food industry has also investigated alternative lethal agents, such as electric fields, high pressure, irradiation, etc., to control microorganisms. Even though it is desirable that the preservation method by itself does not cause any damage to the food, depending upon the intensity of such agents, the quality of the food may also be affected.

## Processes using Addition or Removal of Heat

### Pasteurization and Blanching

Thermal pasteurization (named after inventor Louis Pasteur) is a relatively mild heat treatment, in which liquids, semi-liquids or liquids with particulates are heated at a specific temperature (usually below 100 °C) for a stated duration to destroy the most heat-resistant vegetative pathogenic organisms present in the food. This also results in shelf life extension of the treated product. Different temperature-time combinations can be used to achieve pasteurization. For example, in milk pasteurization, heating temperatures vary widely, ranging from low-temperature, long-time heating (LTLT, 63 °C for a minimum of 30 min), to high-temperature, short-time heating (HTST, 72 °C for a minimum of 15 sec), to ultra-pasteurization (135 °C or higher for 2 sec to 2 min). In addition to destruction of pathogenic and spoilage microorganisms, pasteurization also achieves almost complete destruction of undesirable enzymes, such as lipase in milk. In recent years, the term "pasteurization" is also extended to destroying pathogenic microorganisms in solid foods (such as pasteurization of almonds through oil roasting, dry roasting, and steam processing).

The intensity of thermal treatment needed for a given product is also influenced by product pH; for example, fruit juices (pH <4.5) are generally pasteurized at 65 °C for 30 min, compared to other low-acid vegetables that need to be treated at 121 °C for 20–30 min. As a moderate heat treatment, pasteurization generally causes minimal changes in the sensory properties of foods with limited shelf life extension. Further, pasteurized products require refrigeration as a secondary barrier for microbiological protection.

Blanching, a mild thermal treatment similar in temperature-time intensity to pasteurization is applied to fruit and vegetables to primarily inactivate enzymes that catalyze degradation reactions. This treatment also destroys some microorganisms. It is achieved by using boiling water or steam for a short period of time, 5–15 min or so, depending on the product. Other beneficial effects are color improvement and reducing discoloration. Blanching is often used as a pre-treatment to thermal sterilization, dehydration, and freezing to control enzymes present in the food. Other benefits of blanching include removal of air from food tissue and softening plant tissue to facilitate packaging into food containers.

### Thermal Sterilization

Thermal sterilization involves heating the food to a sufficiently high temperature (>100 °C) and holding the product at this temperature for a specified duration, with the goal of inactivating bacterial spores of public health significance. This is also known as canning or retorting. Prolonged thermal exposure during heating and cooling can substantially degrade product sensory and nutritional quality. Commercial sterility of thermally processed food, as defined by the US Food and Drug Administration (FDA), is the condition achieved by the application of heat that renders the food free of (i) microorganisms capable of reproducing in the food under normal non-refrigerated storage and distribution conditions, and (ii) viable microbial cells or spores of public health significance. Consequently, commercially sterile food may contain a small number of viable, but dormant, non-pathogenic bacterial spores. Traditionally, food processors use severe heat treatment to eliminate 12-log of °C. botulinum spores (i.e. 12-D processes) to sterilize low-acid (pH ≥4.6) canned foods. Many canned foods have shelf lives of 2 years or longer at ambient storage conditions.

## Aseptic Processing

Aseptic processing, a continuous thermal process, involves pumping of pumpable food material through a set of heat exchangers where the product is rapidly heated under pressure to ≥130 °C to produce shelf-stable foods. The heated product is then passed through a holding tube, wherein the temperature of the product mixture is equilibrated and held constant for a short period as determined by the type of food and microbes present, and passes through set of cooling heat exchangers to cool the product. The sterilized cooled product is then aseptically packaged in a pre-sterilized package. Conventional aseptic processing technologies utilize heat exchangers such as scraped surface heat exchangers. Advanced food preservation techniques may utilize ohmic heating or microwave heating instead.

## Sous-vide Cooking

Sous-vide cooking involves vacuum packaging food before application of low-temperature (65–95 °C) heating and storing under refrigerated conditions (0–3 °C). Meat, ready meals, fish stews, fillet of salmon, etc. are some examples of sous-vide cooked products. This technology is particularly appealing to the food service industry, and has been adopted mainly in Europe. Due to use of modest temperatures, sous-vide cooking is not lethal enough to inactivate harmful bacterial spores. In addition, vacuum packaging conditions could also support potential survival of Clostridium botulinum spores.

## Microwave Heating

Microwave energy (300–300,000 MHz) generates heat in dielectric materials such as foods through dipole rotation and/or ionic polarization. In microwave heating, rapid volumetric heating could reduce the time required to achieve the desired temperature, thus reducing the cumulative thermal treatment time and better preserving the thermo labile food constituents. A household microwave oven uses the 2450 MHz frequency for microwave. For industrial application, a lower frequency of 915 MHz is selected for greater penetration depth. Microwave heating can be operated in both batch and continuous (aseptic) operations. Care must be taken to avoid non-uniform heating and overheating around the edges. In 2010, the FDA accepted an industrial petition for microwave processing of sweet potato puree that is aseptically packaged in sterile flexible pouches.

## Ohmic heating

Ohmic heating involves electrical resistance heating of the pumpable food to rapidly heat the food material. The heat is generated in the form of an internal energy transformation (from electric to thermal) within the material as a function of an applied electric field (<100 V/cm) and the electrical conductivity of the food. Ohmic heating has been shown to be remarkably rapid and relatively spatially uniform in comparison with other electrical methods. Therefore, the principal interest has traditionally been in sterilization of those foods (such a high-viscosity or particulate foods) that would be difficult to process using conventional heat exchange methods. Another application of ohmic heating includes improvement of extraction, expression, drying, fermentation, blanching, and peeling.

## Drying

Drying is one of the oldest methods of preserving food. The spoilage microorganisms are unable to grow and multiply in drier environments for lack of free water. Drying is a process of mobilizing the water present in the internal food matrix to its surface and then removing the surface water by evaporation. Drying often involves simultaneous heat and mass transfer. Most drying operations involve changing free water present within the food to vapor form and removing it by passing hot air over the product.

During drying, the heat is transferred from an external heating medium into the food. The moisture within the food moves towards the surface of the material due to the vapor pressure gradient between the surface and interior of the product. The moisture is then evaporated into the heat transfer medium (usually air). The heat transfer can be accomplished through conduction, convection or radiation. While convective heat transfer is the dominant mechanism at the surface, heat is transferred through conduction within the food material. The moisture movement within the food material utilizes a diffusion process. There are several drying and dehydration methods frequently used in food processing such as hot air drying, spray drying, vacuum drying, freeze drying, osmotic dehydration, etc. During hot air drying, heat is transferred through the food either from heated air or from heated surfaces. Vacuum drying involves evaporation of water under vacuum or reduced pressures. Freeze drying involves removing the water vapor through a process called sublimation. Freeze drying helps to maintain food structure.

# Refrigeration and Freezing

Refrigeration and freezing have become an essential part of the food chain; depending on the type of product, they are used in all stages of the chain, from food processing, to distribution, retail, and final consumption at home. These two unit operations take away heat energy from food systems and maintain the lower temperatures throughout the storage period to slow down biochemical reactions that lead to deterioration. The food industry employs both refrigeration and freezing processes where the food is cooled from ambient to temperatures above 0 °C in the former and between −18 °C and −35 °C in the latter to slow the physical, microbiological, and chemical activities that cause deterioration in foods.

Chilled or refrigerated storage refers to holding food below ambient temperature and above freezing, generally in the range of −2 to ~16 °C. Removing sensible heat energy from the product using mechanical refrigeration or cryogenic systems lowers the product temperature. Many raw products (such as milk and poultry) are rapidly chilled prior to further processing to minimize any microbial growth in the raw product. After cooking, foods are often kept under refrigerated conditions during storage and retailing. Many of the minimally processed foods (e.g. pasteurization) are promptly refrigerated to prevent growth of the microorganisms that survive processing.

For frozen storage, food products are frozen to temperatures ranging from −12 °C to −18 °C. Appropriate temperature control is important in freezing to minimize quality changes, ice recrystallization, and microbial growth. Food products can be frozen using either indirect contact or direct contact systems. In indirect contact systems, there is no direct contact between the product and the freezing medium. Cold air and liquid refrigerants are examples of freezing media used. Cabinet freezing, plate freezing, scraped surface heat exchanger, and indirect contact air-blast systems are different examples of indirect freezing equipment used in the industry. Direct contact freezing systems do not have a barrier between the product and the freezing medium. Direct contact air-blast, fluidized bed, immersion freezing, and spiral conveyor systems are examples of direct contact freezing.

# Non-thermal Food Processing and Preservation

During the past two decades, due to increased consumer interest in minimally processed foods with reduced preservatives, several non-thermal preservation methods have been investigated. These technologies often utilize lethal agents (such as pressure, irradiation, pulsed electric field, ultraviolet irradiation, and ultrasound, among others) with or without combination of heat to inactivate microorganisms. This helps to reduce the severity of thermal exposure and preserve product quality and nutrients. Since the mechanism of microorganism inactivation by non-thermal lethal agents may be different from that of heat, it is important to understand the synergy, additive, or antagonistic effects of sequential or simultaneous combinations of different lethal agents.

Irradiation, high-pressure processing, and pulsed electric field processing are examples of nonthermal processing methods that may be of commercial interest.

## Irradiation

Irradiation is one of the most extensively investigated non-thermal technologies. Ionizing radiation includes γ-ray and electron beam. During irradiation of foods, ionizing radiation penetrates a food and energy is absorbed. Absorbed dose of radiation is expressed in grays (Gy), where 1 Gy is equal to an absorbed energy of 1 J/kg. Milder doses 0.1–3 kGy, called "radurization," are used for shelf life extension, control of ripening, and inhibition of sprouting. Radicidation is carried out to reduce viable non-spore forming pathogenic bacteria, using a dose between 3 and 10 kGy. Radappertization from 10 kGy to 50 kGy enables the sterilization of bacterial spores. From its beginning in the 1960s, the symbol Radura has been used to indicate ionizing radiation treatment.

Radiation is quite effective in penetrating through various packaging materials. However, the radiation dose may cause changes in packaging polymers. Thus, careful choice of packaging material is critical to avoid any radiolytic products from packaging contaminating the food products. Consumer acceptance is one of the barriers to widespread adoption of irradiation for food processing applications.

## High-pressure Processing

International Radura symbol for irradiation on the packaging of irradiated foods.

High-pressure processing, also referred to as "high hydrostatic pressure processing" or "ultra-high pressure processing," uses elevated pressures (up to 600 MPa), with or without the addition of external heat (up to 120 C), to achieve microbial inactivation or to alter food attributes. Pressure pasteurization treatment (400–600 MPa at chilled or ambient conditions), in general, has limited effects on nutrition, color, and similar quality attributes. Uniform compression heating and expansion cooling on decompression help to reduce the severity of thermal effects such as quality degradation and nutritional loss encountered with conventional processing techniques. Figure summarizes typical pressure and temperature levels for various food process operations.

Examples of high pressure pasteurized products commercially available in the US include smoothies, guacamole, deli meat slices, juices, ready meal components, poultry prod-ucts, oysters, ham, fruit juices, and salsa.

Heat, in combination with pressure, is required for spore inactivation. This process is called pressure-assisted thermal processing (PATP) or pressure-assisted thermal sterilization (PATS). During PATP, preheated (70–85 °C) food material is subjected to a combination of elevated pressure (500–700 MPa) and temperature (90–120 °C) for a specified holding time. PATP has shown better preservation of textural qualities in low-acid vegetable products. Minimal thermal exposure with a shorter pressure holding time helps to retain product textural quality attributes in comparison with conventional retort processing where the product experiences prolonged thermal exposure. In 2010, the FDA issued no objection to an industrial petition for sterilizing low-acid mashed potato product through pressure-thermal sterilization.

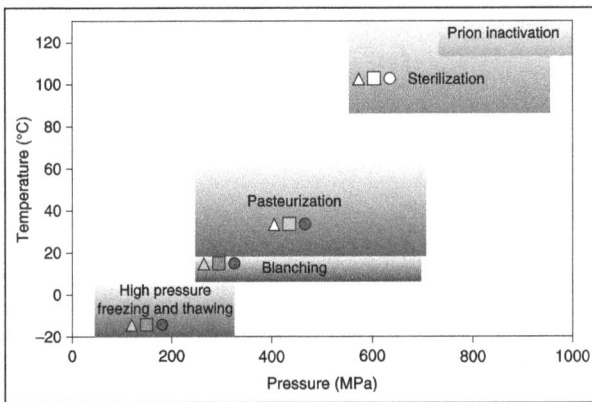

Different pressuretemperature regions yield different processing effects. Inactivation of vegetative bacteria, yeast, and mold (□), bacterial spores (o), and enzymes (Δ) is also shown. A filled symbol represents no effect, an open symbol represents inactivation.

## Pulsed Electric Field Processing

During pulsed electric field (PEF) processing, a highvoltage electrical field (20–70 kV/cm) is applied across the food for a few microseconds. A number of process parameters including electric field strength, treatment temperature, flow rate or treatment time, pulse shape, pulse width, frequency, and pulse polarity govern the microbiological safety of the processed foods. Food composition, pH, and electrical conductivity are parameters of importance to PEF processing. During PEF treatment, the temperature of the treated foods increases due to a electrical resistance heating effect. This temperature increase can also contribute to the inactivation of microorganisms and other food quality attributes. The technology effectively kills a variety of vegetative bacteria, but spores are not inactivated at ambient temperatures. Typical PEF equipment components include pulse generators, treatment chambers, and fluid-handling systems, as well as monitoring and control devices. PEF technology has the potential to pasteurize a variety of liquid foods including fruit juices, soups, milk, and other beverages.

## Ultrasound

High-power ultrasound processing or sonication is another alternative technology that has shown promise in the food industry, especially for liquid foods, in inactivating spoilage microorganisms. Ultrasound is a form of energy generated by sound waves of frequencies above 16 kHz; when these waves propagate through a medium, compressions and depressions of the medium particles create micro-bubbles, which collapse (cavitation) and result in extreme shear forces that disintegrate biological materials. Sonication alone is not very efficient in killing bacteria in food, as this would need an enormous amount of ultrasound energy; however, the use of ultrasound coupled with pressure and/or heat is promising.

## Redefining Pasteurization

Successful commercial introduction of a number of non-thermal pasteurized products prompted the National Advisory Committee on Microbiological Criteria for Foods (NACMCF) to suggest a new definition for pasteurization (National Advisory Committee on Microbiological Criteria for Foods, 2006). According to NACMCF recommendations, pasteurization is defined as "any process, treatment, or combination thereof, that is applied to food to reduce the most resistant microorganism(s) of public health significance to a level that is not likely to present a public health risk under normal conditions of distribution and storage." High-pressure and PEF processing, ultraviolet processing, γ-irradiation, and other non-thermal processes are examples of processes that potentially satisfy the new definition of pasteurization.

## Other Food Processing/Preservation Technologies

### Fermentation

Fermentation causes desirable biochemical changes in foods in terms of nutrition or digestion, or makes them safer or tastier through microbial or enzyme manipulations. Examples of fermented foods are cheese, yogurt, most alcoholic beverages, salami, beer, and pickles. Representative vegetative bacteria in the fermentations are Lactobacillus, Lactococcus, Bacillus, Streptococcus, and Pseudomonas spp. Yeast and fungi (e.g. Saccharomyces, Endomycopsis, and Monascus) are also used for fermentation. Food supports controlled growth of these microorganisms, which modify food properties (texture, flavor, taste, color, etc.) via enzyme secretion.

### Extrusion

Extrusion is a process that converts raw material into a product with a desired shape and form, such as pasta, snacks, textured vegetable protein, and ready-to-eat cereals, by forcing the material through a small opening using pressure. Some of the unique advantages of extrusion include high productivity, adaptability, process scale-up, energy efficiency, low cost, and zero effluents. An extruder consists of a tightly fitting screw rotating within a stationary barrel. Within the extruder, thermal and shear energies are applied to a raw food material to transform it to the final extruded product. Preground and conditioned ingredients enter the barrel where they are conveyed, mixed, and heated by a variety of screw and barrel configurations. Inside the extruder, the food may be subjected to several unit operations, including fluid flow, heat transfer, mixing, shearing, size reduction, and melting. The product exits the extruder through a die, where it usually puffs (if extruded at >100 °C and higher than atmospheric pressure) and changes texture from the release of steam and normal forces. Extruded products may undergo a number of structural, chemical, and nutritional changes including starch gelatinization, protein denaturation, lipid oxidation, degradation of vitamins, and formation of flavors. Very limited studies are available to describe kinetics changes in foods during extrusion.

## Baking

Baking uses dry heat to cook fully developed flour dough into a variety of baked products including bread, cake, pastries, pies, cookies, scones, crackers, and pretzels. The dough needs to undergo various stages (mixing, fermentation, punching/sheeting, panning, proofing among others) before it is ready for baking.

Carbon dioxide gas is produced from yeast fermentation of available sugars, which could either be added or obtained via amylase breakdown of starch. During baking, heat from the source in an oven is transferred to the dough surface by convection; from the surface, it then transfers via conduction. As the heat is conducted through the food, it transforms the batter or dough into a baked food with a firm crust and softer center. During baking, heat causes the water to vaporize into steam. Gelatinization of flour starch in the presence of water occurs. The protein network (gluten) holds the structure, while carbon dioxide gas, that gives the dough its rise, collapses during baking (at ~450 °C). The product increases in size and volume, called leavening. Baking temperatures also cause a number of biochemical changes in the batter and dough, including dissolving sugar crystals, denaturing egg and gluten proteins, and gelatinizing starch. Baking also causes the surface to lose water, and breaks down sugars and proteins on the surface of the baked goods. This leads to formation of a brown color and desired baked flavor.

## Hurdle Technology

Hurdle technology involves a suitable combination of different lethal agents to ensure microbial safety, quality, and stability of the processed product. Heat, pressure, acidity, water activity, chemical/natural preservatives, and packaging are examples of hurdles that can be combined to improve the quality of the final processed product. The hurdle approach requires the intensity of individual lethal agents (for example, heat or pressure) to be relatively modest, yet is quite effective in controlling microbial risk. Efforts must be made to understand the potential synergistic, additive or antagonistic effects of combining different lethal agents during hurdle technology.

## Packaging

Packaging plays a vital role in many food preservation operations. Packaging has many functions, including containment, preservation, communication/education, handling/ transportation, and marketing. Packaging helps maintain during storage the quality and properties of foods attained via processing. The packaging protects the food material from microbiological contaminants and other environmental factors. The package also helps prevents light-induced changes in stored food products and minimizes loss of moisture.

Depending upon the intensity of lethal treatments (heat, pressure, radiation dose), processing not only affects the food material but also alters the (moisture and oxygen) barrier properties of packaging materials and possibly induces migration of polymer material into the food. Thus, careful choice of food packaging material is essential for successful food process operation.

## Emerging Issues and Sustainability in Food Processing

Modern food processing was developed during the 19th and 20th centuries with the rise of thermal pasteurization and sterilization techniques with the view of the extending shelf life of processed foods. Developments in industrial food processing technologies ensured the availability of a safe, abundant, convenient food supply at reasonable prices. However, the industry is currently undergoing a transformation in response to a variety of societal challenges. In a recent IFT scientific review, Floros et al. identified three emerging societal issues that will likely shape future developments in food processing.

- Feeding the world: The world population today is about 6.8 billion and it is expected to reach about 9 billion by 2050. Sustainable and efficient industrial food processing technologies at reasonable cost are needed to feed this ever-expanding world population.

- Overcoming negative perceptions about "processed foods:" There has been an increased negative perception towards processed foods in the US. A number of societal factors have contributed to this trend, including negative perceptions towards technology use, diminishing appreciation of scientific literacy, as well as lack of familiarity or appreciation about farming among increasingly urban based consumers.

- Obesity: Overweight and obesity are major health problems in the US and developed countries. Overconsumption of calorie-dense processed food and sedentary consumer lifestyles are some reasons put forward for the increase in obesity.

Apart from these issues, sustainability in the food processing industry is another emerging key societal issue. Sustainability is the capacity to endure; it is utilizing natural resources so that they are not depleted or permanently damaged. Water, land, energy, air, etc. are resources utilized in agriculture and food processing. Environmental concerns related to food production and processing which require consideration include land use change and reduction in biodiversity, aquatic eutrophication by nitrogenous factors and phosphorus, climate change, water shortages, ecotoxicity, and human effects of pesticides, among others. Sustainable food processing technologies emphasize the efficient use of energy, innovative or alternative sources of energy, less environmental pollution, minimal use of water, and recycling of these resources as much as possible. Sustainable food processing requires processors to maximize the conversion of raw materials into consumer products by minimizing postharvest losses and efficient use of energy and water. Modern food processing plants can contribute to sustainability by utilizing green building materials and practices in their construction, utilizing innovative building designs, using energy-efficient equipment and components, and following efficient practices in routing, storing, and processing of ingredients and distribution and handling of finished products.

# Management of Food Quality and Safety

Food quality can be defined as a total of traits and criteria which characterize food as regards its nutritional value, sensory value, convenience as well as safety for a consumer's health. Thus, it is a broader concept than food safety. Food safety (hazard-free) is the most important feature of food quality, hence the food law regulates this issue, in order to assure consumers that the food they purchase meet their expectations as regards safety. It is also an increasingly important public health issue. Governments all over the world are intensifying efforts to improve food safety in response to an increasing number of problems and growing consumer concerns as regards various food risks.

Besides, it is important to distinguish between the terms food quality and food health quality. As figure demonstrates, these two remain in the relationship such as food health quality embracing only the health-related traits (that is, hazard-free and nutritional value), whilst food quality being a broader concept, covering all the features presented. Thus, in addition to food health quality related attributes, food quality comprises values such as sensory characteristics (e.g. taste of food, smell, etc.) and convenience (e.g. easy in preparation, etc.).

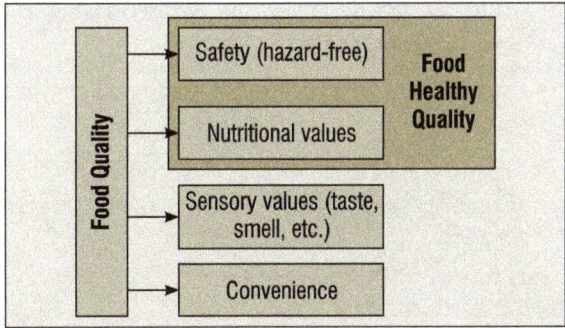

Diagram of the relationship between food quality, food health quality and food safety traits.

In order to preserve the above quality features in food products, various safety and quality assurance systems have been developed. Any system constitutes a systematic approach to assure that food products have particular traits at any stage of production and distribution. Some of the systems are obligatory by law and some voluntary to be implemented by the food chain members.

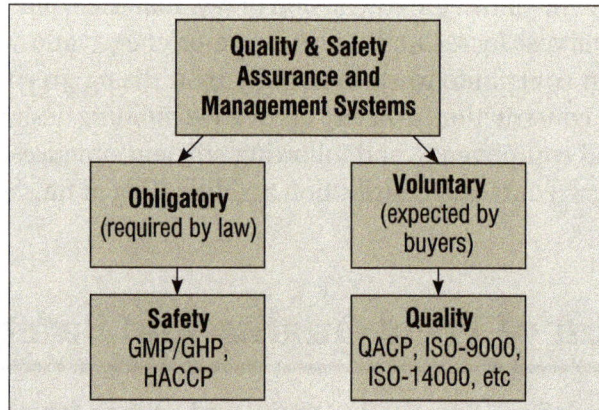

Diagram of voluntary vs. obligatory quality and safety systems.

## Safety Assurance Systems

The distinction between obligatory and voluntary systems is based on the safety (hazard-free products) being the quality of food required by law. Thus, obligatory systems have been established to assure food safety, and are subsequently called "safety assurance systems". These include Good Hygiene Practices (GHP), Good Manufacturing Practices (GMP) and Hazard Analysis and Critical Control Point (HACCP).

Good Manufacturing Practices (GMP) is a set of guidelines specifying activities to be undertaken and conditions to be fulfilled in food manufacturing processes in order to assure that the food produced meets the standards of food safety. Similarly, Good Hygienic Practices (GHP) constitute a set of guidelines specifying activities to be undertaken and hygienic conditions to be fulfilled and monitored at all steps of the food chain in order to assure food safety. Both GMP and GHP constitute a precondition in a food enterprise for implementing the HACCP system. Figure illustrates the relationship between these three food safety assurance systems, where HACCP is a broader category which incorporates its prerequisites GMP and GHP.

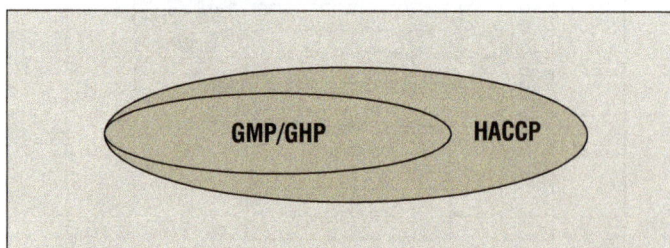

Diagram of the relationship between GMP, GHP and HACCP.

## Quality Assurance and Management Systems

Maintenance and/or introduction of the remaining qualities in food (nutritional, sensory and convenience values), is not requested by law, albeit desirable by customers. Voluntarily implemented systems, known as quality assurance and management systems, include for example Quality Assurance Control Points (QACP), the well-known ISO-9000 (quality management) and ISO-14000 (environmental management).

Quality Assurance Control Points (QACP) is one of the quality assurance systems in food production, created based on the HACCP concept. In case of HACCP, Critical Control Points (eliminating hazards), parameters and their critical limits are determined, while in QACP – Control Points (quality assurance, not safety), parameters and their critical values. Likewise in HACCP, QACP is unique for each company and must be introduced individually for each enterprise and production line.

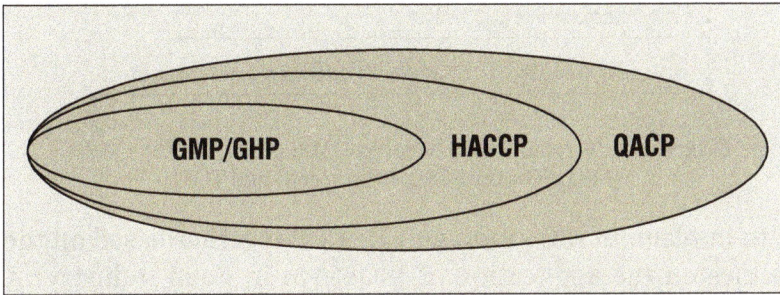

Diagram of the relationship between GMP, GHP, HACCP and QACP.

Having implemented GMP/GHP, HACCP and QACP systems, the next step could be to implement other quality systems, e.g. ISO-9000. The ISO-9000 series of standards represent the requirements which have to be addressed by every enterprise to assure the reliable production and timely delivery of goods and services to the marketplace. There is a body of literature and plenty of professional materials (both electronic and in print) concerning the quality management based on the ISO-9000 series of standards. Many food chain actors require their suppliers to become registered to ISO-9000 and because of that, those who registered find that their market opportunities have increased. Nevertheless, despite being very popular, ISO- series are not, and are not going to be, obligatory.

Figure below illustrates the full range of the safety and quality assurance and management systems and the relationship between them. Considering the above, it is important to make a distinction between the terms "assurance" and "management". The term "assurance" relates to a product itself and involves all the safety assurance systems (GMP, GHP and HACCP) and the quality assurance system QACP. On the other hand, the term "management" corresponds to a company's overall organisation as regards the products' quality (including safety), and involves the remaining Quality Management Systems QMS (ISO-9000, ISO-14000, etc.) as well as Total Quality Management TQM.

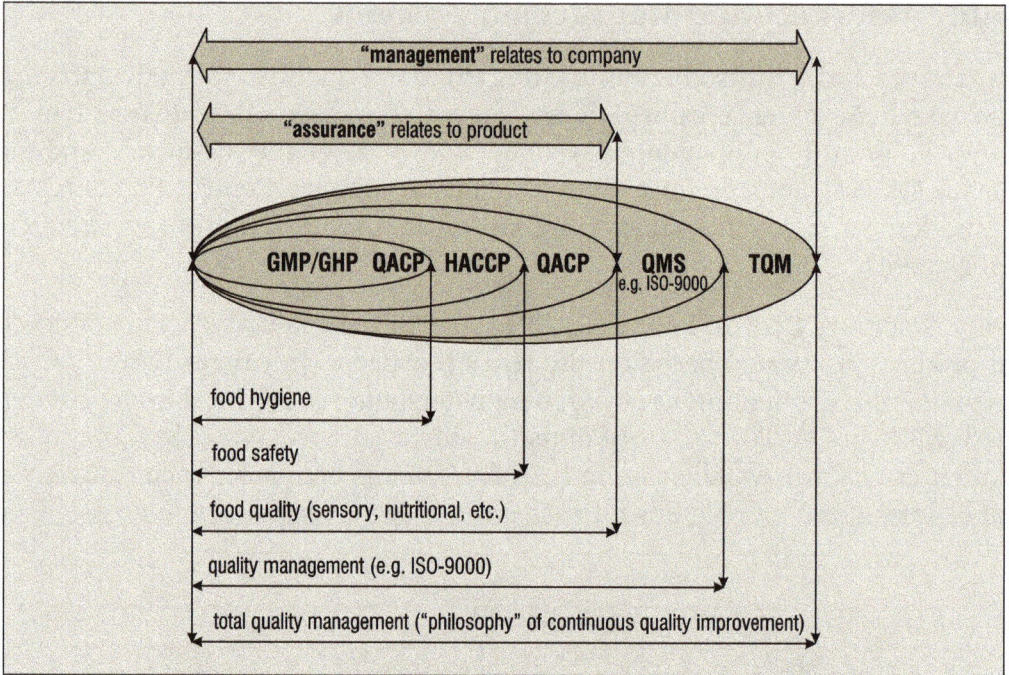

Diagram of the relationship between GMP, GHP, HACCP, QACP,
QMS (ISO-9000, ISO-14000, etc.) and TQM.

It is possible to implement ISO-9000 and HACCP together based on the norm ISO-15161 (Guidelines on the application of ISO-9001 in Food Industry). Currently, a new norm is being prepared (ISO/CD 22000) to internationally harmonize the requirements of food safety management in food industry. Popularity of food safety systems (GMP/GHP, HACCP) and quality systems (e.g. ISO-9000) contributed to the development of traceability systems. A need for traceability emerged also from the growing consumers' concerns as regards food safety. In brief, traceability is a system allowing "tracing" the history of a product at all steps of the food chain. It refers to the origins of raw materials as well as the history of production and distribution of the product.

## Certification

Holding a certificate confirming the quality of a company's products is not a legal requirement. Yet, having a certificate issued by an external institution (such as certifying firm) is part of the marketing strategy, shaping a positive view of a company in the eyes of its customers. This is with no doubt contributing to an increase in market opportunities and consequently, higher profits. In some cases, food enterprises apply for a certificate just because it is required by their clients, trading companies.

Certifying organizations have a very specific status. Presumably, all food companies implementing ISO-9000 would apply for certificate in one of the certifying organizations, while HACCP might be less popular due to the fact that the system has to be implemented obligatorily. A certificate is usually granted for 3 years and paid for by the company who applied for it. Thus, the "quality" of certificates may differ dependent upon the reliability of certifying organization. Besides, a food company which realizes the fact that holding a certificate would not influence its sales would not decide to continue, after 3 years would give up certification whilst maintaining the system.

## Making the Systems Work

As presented in figure below, an important feature which distinguishes food safety and food quality is that the former is regulated by law, while the latter is demanded by those who purchase food (e.g. distributors and/or consumers). Consequently, food safety assurance systems are implemented obligatorily, whilst quality assurance and management systems are voluntary, thus the decision whether to introduce them or not is to be taken by the food chain actors individually. Figure summarizes the relationship between food legislation, safety and quality systems, official inspections and customers' requirements. The combination of these elements forms a machinery of food safety and quality assurance and management.

Food law represents a set of legal norms regarding the principles of production and distribution of raw materials, foodstuffs and objects getting in direct contact with them, to the level which ensures consumer health protection and fulfil consumer expectations as regards food safety. Due to the fact that in the European countries, food law is implemented through the safety assurance systems exclusively (namely GMP, GHP and HACCP), there is no direct link between food law and food safety per se. Both the EU law and national law regulate this issue via the systems. Hence, their implementation needs to be officially supervised by the state inspections, whose activity is also regulated by law. Figure also draws a link between official inspections and quality assurance systems. Despite being implemented voluntarily, food quality may also become a subject for official inspections in terms of conformity between the actual qualities of the product and those declared by a food chain actor. The same applies to food safety. On the other hand, customers may expect food chain actors to have implemented quality assurance and management systems (confirmed by a certificate) not just because of their expectations as regards the product itself, yet being concerned whether the company conducts environmentally friendly practices (e.g. according to ISO-14000 standards).

The fact that the customers' expectations influence the quality and safety of food products seems obvious in that each producer should be concerned about his own reputation at the marketplace. Selling low quality products or unsafe food appears suicidal to a company who would conduct such practices. Unfortunately, such instances do occur, therefore other measures are required to protect the consumers' health and assure that food products meet their expectations.

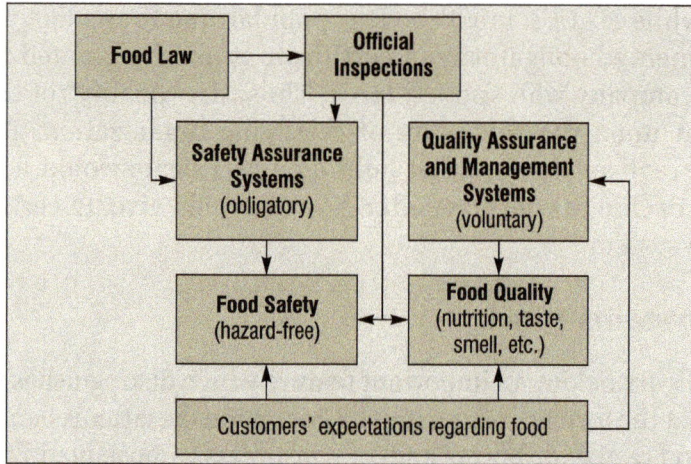

Integrated diagram of the role of food law, official inspections and consumers'
expectations as regards food quality and safety.

Table: The european union food law main directives and regulations.

| European Union Food Law embraces the following: |
| --- |
| • 93/43/EEC Council Directive on the hygiene of foodstuffs; |
| • Regulation 178/2002 defining general food safety rules and establishing of EFSA; |
| • Regulation 852/2004 on the hygiene of foodstuffs to replace 93/43/EEC Council Directive; |
| • Regulation 853/2004 on the hygiene of foods of animal origin; |
| • Regulation 854/2004 determining the control points of foods of animal origin. |

It is necessary to remember, however, that no system can work without ethics. The
role of ethics in the food industry was recently brought up in Poland, following a food
scandal in a well known meat plant, which was suspected to be conducting unsanitary
practices. When supermarkets returned deli meats after they expired, the plant did
not destroy them as required, but they were reprocessed, re-packed and re-sold to the
stores. Despite frequent sanitary inspections, these practices persisted until the un-
dercover TV reporter (who took a job as a plant worker and subsequently documented
these practices) made the discovery and unveiled it to the public.

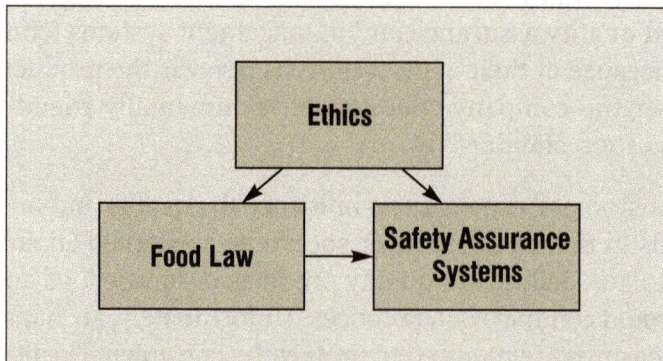

Diagram of the role of ethics in food law and food safety.

# Food Control

Food control plays an important role in assuring high quality, safe, and nutritious food for the public, for their good health, and for the economic benefits derived from trade in safe and high-quality food products.

Although food control has been described in many different ways, the Food and Agriculture Organization of the UN (FAO) adopted a definition that appears to be most suitable. It is defined as "a mandatory regulatory activity of enforcement by national or local authorities to provide consumer protection and ensure that all foods during production, handling, storage, processing, and distribution are safe, wholesome, and fit for human consumption; conform to the quality and safety requirements; and, are honestly and accurately represented in its labeling as prescribed by law."

Official food control has its foundation in law. Any effort to revise food law should incorporate the input of the affected sectors, such as the food industry and consumers. Effective control requires commitment and resources. The laboratory function is critical to food control. Compliance functions in food control vary from country to country. Scientific advisory services are absolutely necessary for a food control system. Using a risked-based programmed approach to food control requires understanding food hazards and how to control, reduce, or eliminate these hazards to decrease consumer health risks. Consumer Affairs is another important aspect of food control.

Most countries need to update their food laws and regulations to reflect rapidly developing changes coming about from the implementation of modern food control methods. Many countries place their food control emphasis in laboratory analysis of finished food products, consequently, laboratory personnel have graduate level university degrees.

The Impact of International Food Trade is increasing as the volumes of food moving across international borders increases. The World Trade Organization's (WTO) statistics indicate international trade in food to be more than US$400 billion dollars. The World Trade Organization (WTO) has had a dramatic impact on the food trade. As it relates specifically to food, it includes the control of food additives, food contaminants, toxins, and disease-causing organisms.

Since 1962, the Codex Alimentarius Commission (CAC) has been responsible for implementing the Joint FAO/WHO Food Standards Program. The name Codex Alimentarius is taken from Latin and translates literally as food code or food law. The Commission's primary objectives are the protection of the health of consumers, the assurance of fair practices in the food trade, and the coordination of all food standards work. This action created a new emphasis and importance being placed on the work of the Codex in establishing international food quality and safety standards. The role of Codex in establishing quality and safety standards for food in international trade become exceptionally more important.

At the international level, there are some dramatic changes taking place in food control, somewhat driven from the changes being made in the international food trade. There has been a reorientation in the direction of food control. For example, in the Western nations, single food safety agencies are being considered.

The use of food standards to protect consumers and facilitate trade has a long tradition. Rules were laid down by Moses to prevent the consumption of meat from unclean animals, especially animals that had died from causes other than supervised slaughter. Assyrian writing tablets provided descriptions on how one goes about determining the correct weights and measures for food grains. In the early history of Athens, beer and wines were inspected to ensure the purity and soundness of these products. The Romans provided a well-organized state-controlled food control system to protect consumers from fraud by preventing the sale of bad quality or adulterated produce.

In the European Middle Ages, individual countries passed laws concerning the quality and safety of eggs, sausages, cheese, beer, wine, and bread. Some of these statutes are still in use today. The Industrial Revolution gave technological and economic impetus to trade in foods across and between continents. Food chemistry, as a science, dates from the mid to late 1800s, and most early food standards and the basic startup food control systems also date from about this same time.

Since the early 1900s comprehensive food laws providing for food standards and other consumer protection measures have continued to evolve and expand to provide a greater degree of protection and to deal with the increased levels of sophistication of the problems associated with food. This was particularly true following the introduction of mass production technology and to the emergence of larger urban centers in the late eighteenth century and early nineteenth century. New and different approaches to food control and consumer protection were required to address public concerns for food problems emerging from these dramatic changes. The response of most countries undergoing these changes was to enact food laws and regulations, and to establish official organizations and institutions to administer food control activities. This approach laid the foundation and became the forerunner of today's food control system.

## Elements of a Food Control System

Although food control has been described in many different ways, the Food and Agriculture Organization of the United Nations (FAO) adopted a definition that appears to be most suitable. It is defined as "a mandatory regulatory activity of enforcement by national or local authorities to provide consumer protection and ensure that all foods during production, handling, storage processing, and distribution are safe, wholesome and fit for human consumption; conform to the quality and safety requirements; and, are honestly and accurately represented in its labeling as prescribed by law." This definition was published by FAO in the Food and Nutrition Paper (FNP) series on Food Quality Control; Manual 14/11 entitled *Management of Food Control Programs.*

Official food control has its foundation in law. Without legal authority as indicated by a legal instrument of the government, there is no authority for official activities and no creditability for those carrying out the activity.

Therefore, to begin improving a food control system, the legal framework on which food control is based, should be reviewed, revised where necessary, and updated to meet the current environment. Food law is an expression of the will of the government to assure food quality and safety and to carry out the required consumer protection measures as a matter of public policy. Food law should provide clear definition of the terms used in the law. It establishes the procedures to administer the law, including the authority to promulgate rules, regulations, codes of practices, quality, and safety standards, and procedures for food handling, processing, storage, shipping, and sale. It should define the role and authority of the competent agency of government and the powers granted to the agency and to the personnel of the agency. It should define the role and responsibility of the private sector and other institutions where applicable, such as industry, academic institutions, scientific committees and consumers. Any effort to revise food law should incorporate the input of the effected sectors, such as the food industry and consumers.

General regulations issued under authority of the law should be specific and clear and in plain language so as to be understood by nontechnical people as much as possible. It is useful to obtain the input of those affected by the regulations to consider before issuing the final rules. There are occasions when the agency may issue regulations, which are sensitive from a regulatory point of view, and these may issue without outside agency input. In either case, they should clearly state the requirements, limits, or other restrictions. It is a good policy to follow up the issuance of any regulation with information material designed to provide explanations and answers to anticipated questions. When possible, educational programs, training seminars, and workshops can be carried out to facilitate understanding and compliance.

Official food control requires commitment and resources. It is important to obtain this commitment and the resource through developing an appropriate strategy as to what is to be achieved, how it will be done, who will do it, and in what time frame it will happen. The development and implementation of a national strategy is an endeavor that should involve all the interested parties. In many countries, a working committee has been established to make a review of the existing situation in the country related to food control, health, and trade, often with the assistance of an international organization or expert consultants. The committee should be made up of individuals with technical skills (academic, science, legal, and industrial), social/cultural and economic skills, and knowledge from as many of the affected sectors as possible.

The primary functional units of a food control system at the basic and minimal level includes an inspectorate, an analytical service, and a regulatory compliance unit. The Inspectorate conducts inspections and investigations of industry's performance in complying with official control requirements. The Analytical Service performs the testing and examination of products for determination of compliance with mandatory requirements of law and regulations, including mandatory food standards, established quality and safety limits for chemical and biological contaminants, packaging requirements, and other factors for which testing is required. The Compliance Unit serves as the enforcement function to oversee the bringing of legal cases when warranted. Other functional units support these activities and include administrative, planning, programming, research and information, education and training support, to assist both internal agency units and affected external sectors.

The Information and Training Services can conduct workshops and seminars on timely subjects of concern to industry or consumers, or can develop informational materials to be used for public distribution. A science and technology services group provides the support needed in research planning and support, or for review of the latest technologies in food control or food processing, and can liaison with academia on technology transfer, to assist in solving technical problems. Finally, a consumer affairs unit would relate to consumer issues, and work with consumer groups and the media and public in general to describe the food control programs, get input and provide information that is useful and informative. The unit can also assist in dealing with information and press media relations to get important food quality and safety messages to the public, particularly during times when emergency situations require public involvement in protection against health-threatening hazards related to the food supply.

## Inspectorate

A primary functional unit of official food control is an adequately staffed and trained inspectorate. The role of the inspectorate is to inspect domestic food manufacturing, processing, and handling facilities, and import/export foods, and their facilities for compliance with the national legal and regulatory requirements. The inspector normally collects necessary samples for all types of food analysis to demonstrate the compliance level of any suspected foods, and for market samples for monitoring and surveillance purposes. In many countries, the inspectors also conduct investigations of suspected food poisoning or injury, fraudulent marketing and handling practices, complaints by consumers or industry, and illegal importation or exportation of food products.

It is important that the inspectors be trained in the latest investigative techniques, and be fully educated on the latest food safety and quality assurance methods, including the strengths and weakness of each method. They should receive up-to-date training in the new technologies used in processing and manufacturing, including what is required to control these technologies to function at maximum effectiveness and to assure proper technological performance. They must also be able to evaluate the performance of equipment and instruments used in production to assure they are appropriately controlled and monitored. In short, they must be well trained and understand the importance of Good Manufacturing Practices (GMPs), recognize deviations from GMPs, and know the impact on product quality and safety when they occur. And finally, the inspector should understand the utilization and application of Hazard Analysis and Critical Control Point (HACCP) based systems for the added margin of safety needed to the existing quality assurance and control measures used by the food industry.

The laboratory function is critical to food control. The laboratory personnel confirm the suspicion of the inspector that the food products sampled are not in compliance. They also confirm the quality and safety of food by determining if mandatory levels or limits of contaminants, additives, or other restricted materials are met and if the product complies with mandatory food standards. The laboratory is the gathering place of analytical data related to monitoring activities such as for food contaminants, microbiological contamination, meeting quality and safety standards, and so on. Laboratories deal with complex analytical problems caused by product composition interference. The problems can only be overcome by using the latest in analytical instruments and sophisticated methods of analysis. This also requires up-to-date technical knowledge acquired only through a continuous personal training program.

Because the analytical results may serve as one of the aspects for legal action against a food producer, they must therefore be accurate and precise. In the legal system of most countries, these matters come under careful scrutiny, with the laboratory analyst frequently forced to defend his technical abilities, the method and techniques used in the analysis, the accuracy and precision of the instruments used, and finally, the results of analysis. Official food control laboratories must maintain an internal quality assurance program to assure their credibility under such circumstances.

Compliance functions in food control vary from country to country. It is not always the responsibility of the food control authority to carry out court actions, and more commonly it is the responsibility of a legal unit in the Ministry of Justice. However, the food control unit is usually the unit that recommends that penalty actions should be taken based on evidence of violations documented during their inspection or investigation. In some cases, the food control authority may have limited powers to take minor penalty actions without court or legal intervention, commonly referred to as administrative action on the order of the Minister, such as establishment closure, or license suspension. In either case, the food control unit then should have a compliance unit. This unit would assure that the recommendations for legal action to the legal unit meet legal criteria for the matter to be referred onward to the courts and that evidence is available to prove violations. In penalty actions handled directly by the food control authority, the compliance unit would be the agency's legal representative and would deal with the administrative action as required by law and within government policy requirements.

The compliance unit would be responsible for both actions that are considered as regulatory in nature, such as court actions, and for programs intended to achieve compliance through voluntary means. It is an accepted premise in food control that most food establishments and businesses will comply with reasonable rules and laws provided they understand what they must do to comply, and that they believe it is in their best interest to do so. Consequently, keeping the industry informed about the requirements, and working with them to assist in achieving these requirements, will go a long way in assuring food is safe and of suitable quality without having to resort to penalty actions. Some of the supporting functions to the food control activity include information, education, and training services. These functions may not be a direct part of the food control functional unit, but may be in other agencies of government acting in a horizontal manner across government agency lines. In this case, they may have functions in a variety of areas, such as health education, trade, and industry information services, consumer information services, and so on. In either case, it is an important element of the food control process because industry and consumers alike need to have available information to make decisions in the business world and marketplace. Food control officials must recognize the need for the development of information in a useable format that keeps people informed about important aspects of the food supply. Communications through public media announcement, published brochures, information bulletins, even the development of an Internet home page can go a long way in keeping people informed. Programs of education and training can be arranged directly or through educational institutions.

Scientific advisory services are absolutely necessary for a food control system. Using a risked based programmed approach to food control requires understanding food hazards and how to control, reduce, or eliminate these hazards to decrease consumer health risks. The scientific community plays an important role in developing methods, conducting research, and defining the severity of the risk to consumers. They can assist in solving technical problems and provide sound scientific information to support and defend actions taken. They can provide risk assessment estimates on additives to food, contaminants, and residues when necessary, particularly in circumstances when a higher level of protection is needed for the public than international standards provide, or where international standards do not exist. Consumer Affairs is another important aspect of food control. Consumers must understand there is no such thing as an absolutely safe food supply. They must understand how they can protect food in the household, during food handling, preparation, and serving, and when handling and storing food left over to prevent this food from becoming a health hazard. Consumers need a focal point in the food control system to let their dissatisfactions be known, to complain about product deception and poor quality, and to report injury and illness caused by food. The Consumer Affairs support function provides opportunity for food control officials to have a channel directly to consumers in time of emergency and need for public awareness or warning. A trained specialist in public relations and consumer affairs with reasonable technical skills can be an exceptional asset and resource to the food control unit.

# Hygienic Design of Processing Facilities

Increasing consumer demand for fresh foods has led to the development of processing and preservation methods that have minimal impact on either the nutritional or sensory properties of foods. Freshly prepared foods often contain less salt, acid, sugar, additives and preservatives. Since the use of mild preservation technologies primarily results in pasteurized products, hygienic processing equipment and a hygienic process environment are needed to prevent microbial, chemical and physical contaminants from affecting these products while preventing product exposure to sources of filth (pests, dust, etc.). Combating product contamination may occur not only at the equipment level but also at the factory level. Incorporation of hygienic design into your food processing facility can prevent development of pests and microbiological niches; avoid product contamination with chemicals (e.g., cleaning agents, lubricants, peeling paint, etc.) and particles (e.g., glass, dust, iron, etc.); facilitate cleaning and sanitation and preserve hygienic conditions both during and after maintenance. The facility infrastructure can be so designed and constructed that it cannot contaminate food products, whether directly or indirectly.

## Barrier Technology

To control food safety, providing barriers to food contamination is a generally applied concept. The first barrier refers to outside premises, such as fencing, to prevent unauthorized access to the facility. The access of transport vehicles with raw materials and end-products, personnel, domestic and non-domestic animals should be monitored and controlled. Factory site drainage and storm water collection must be sufficient; areas within a 3-m perimeter of the factory must be kept vegetation free to avoid pest breeding and harborage sites; a 10-cm thick concrete curtain wall around the factory foundation at least 60 cm below ground discourages rodents from entering the building; effluent treatment plants and waste disposal units should be sited such that prevailing winds do not blow microbial and dust aerosols into manufacturing areas.

The second barrier concerns the closing of factory buildings. All entrances/exits (i.e., window and door openings, openings for vents, air circulation lines, floor drains, etc.) must be designed for control over access, flow or exit of personnel, raw and finished food products, air, process aids (process water, process steam, food gases, etc.), waste, utilities (plant cooling and heating water, plant steam, compressed air, electricity, etc.) and pests (insects, birds, rodents, etc.). Floor drains must be screened to avoid rats from entering the food plant via sewers; ventilator openings, including vents in the roof, should be screened to prevent the entry of roof rats, insects and birds; gaps at the entrances of electrical conduits, process and utility piping, which are convenient pathways for roof rats, must be closed.

The third barrier is the segregation of restricted areas (zones) within the plant, each of which have different hygienic requirements and controlled access. The fourth barrier is the processing equipment (including storage and conveying systems), which must have an adequate hygienic design and must be closed to protect the food product from external contamination.

## Zoning: A Cornerstone in Prevention of Food Contamination

Zone B: Is an area in which a basic level of hygienic design requirements suffices. It encompasses areas in which products are produced that are not susceptible to contamination or that are protected in their final packages. A B0 zone is the area outside the buildings within the perimeter of the site where the objective is to control or reduce hazards created by unauthorized personnel entry and hazards created by water, dirt, dust and presence of animals. B1 zones include warehouses that store both raw materials and packed processed products, offices, workshops, power supply areas, canteens and redundant buildings/rooms. The objective for a B1 zone is to control or reduce hazards created by birds and pests.

Zone M: Is an area in which a medium level of hygiene suffices. It includes process areas where products are produced that are susceptible to contamination, but where the consumer group is not especially sensitive and where no further microbial growth is possible in the product in the supply chain. In this area, product might be exposed to the environment, during sampling and during the opening of equipment to clear blockages. The objective for zone M is to control or reduce the creation of hazardous sources that can affect an associated area of higher zone classification. Another objective is the protection of the interior of food processing equipment from contamination when exposed to the atmosphere.

Zone H: Applies to an area where the highest level of hygiene is required. A "High Hygiene" room, which, in food processing is the equivalent of a cleanroom, must be completely contained. Zone H is typical for open processing, where even short exposure of product to the atmosphere can result in a food safety hazard. Products and ingredients that are processed or stored and are destined for a highly susceptible consumer group (e.g., infant nutrition), are instant in nature or ready for consumption. They must be handled in a refrigerated supply chain, as they are susceptible to growth of pathogenic microorganisms. The objective for H zones is to control all product contamination hazards and to protect the interior of food processing equipment from exposure to atmosphere. Filtered air must be supplied to this area.

These areas should be limited in size, must have a simple equipment layout to facilitate process, cleaning and maintenance operations and should have utilities located outside. However, investing in an enclosed line that brings barriers very close to the product is more logical than trying to create a complete cleanroom around a partially open line.

Zoning and the establishment of barriers to ensure that product of acceptable hygienic quality is produced should only be applied where their use will help significantly to protect products. Designing the entire factory as a cleanroom is not the purpose of food area segregation to protect both product and consumer. Zoning and barrier technology must be applied in an appropriate and consistent way, thereby avoiding unnecessary investment.

## Construction of Facilities: Appropriate Layout

The layout and design of the food factory must be adapted to the hygienic requirements of a given process, packaging or storage area. The interior of the factory must be designed so that the flow of material, personnel, air and waste can proceed in the right direction. As they become incorporated into food products, raw materials and ingredients should move from the 'dirty' to the 'clean' areas. However, the flow of food waste and discarded outer packaging materials should be in the opposite direction. Before building begins, simulation of the flow of people, materials, products and waste can help the designer determine the most appropriate place for installing the process equipment and where the process and utility piping should enter the process area. Even the simulation of maintenance and cleaning operations can be useful to determine the most appropriate factory layout. Graphical computer-aided design and 3D visualization programs can help in the hygienic design, positioning and routing of processes, process supports and utility systems. These programs allow the observer to "walk through" the facility, seeing the inside of the facility from different angles and locations. To save building and renovation costs, potential problems can be solved before the onset of construction. Additionally, in the development of high hygiene areas, computational fluid dynamics can help simulate and visualize expected airflows.

To meet a possible increase of processing activities within the food plant in the future, the building and its food processing support systems should be designed so they can either be expanded, or another building and/or utilities can be added. Oversizing the main utility systems is a common practice. If possible, the factory should also be made adaptable (i.e., the ability to modify the production area for other manufacturing purposes) and versatile (i.e., the ability to do different things within the same room).

## Construction of Facilities: Pest Prevention

To exclude flooding and the entry of rodents, factories should be built at a higher level than the ground outside. Exterior doors should not open directly into production areas, and windows should be absent from food processing areas. The number of loading docks should be minimal and be 1–1.2 m above ground level. Preferably, outside docks should have an overhanging lip, with smooth and uncluttered surfaces that are sloped slightly away from the building to encourage water run-off. Areas beneath docks should not provide harborages for pests, should be paved and should drain adequately. To provide protection for products and raw materials, docks can be shielded from the elements by roofs or canopies. However, these structures can become a serious sanitation problem due to roosting or nesting of birds. Bird spikes or nets can solve that problem. To prevent the entry of insects, dock openings should be provided with plastic strips or air curtains, and external lighting to illuminate these factory entrances should be placed in locations away from the factory building. Intruding insects can still be attracted and killed within the food factory by strategically positioned ultraviolet (UV) light electric grids or adhesive glue board traps.

## Construction of Facilities: Interior Hygienic Design Construction Materials

Construction materials for equipment and utility piping should be hygienic (smooth, non-absorbent, non-toxic and easily cleanable), chemical-resistant (to product, process chemicals, cleaning and sanitizing agents), physically durable (unbreakable, resistant to steam, moisture, cold, abrasion and chipping) and easy to maintain. Materials used to construct process and utility systems located in the non-food contact area may be of a lower grade than those applied in the food contact zone. Surfaces that are frequently wet should not be painted as the paint can crack, flake and chip.

Lead, mercury and cadmium should not be used within the factory. However, as part of many electric components, it is very difficult to exclude their presence. In the food contact area, electric components must always be enclosed in junction boxes, casings, closed cable housings, cabinets, etc. or should be installed in non-product contact zones or in technical corridors. Alloys for food contact may only contain aluminium, chromium, copper, gold, iron, molybdenum, nickel, platinum, silver, titanium, zinc, carbon, etc. However, zinc, copper, aluminium, bronze, brass, carbon and galvanized and painted steel have poor resistance to detergents, disinfectants, acidic food and steam and must be avoided in food contact areas.

Polytetrafluoroethylene, polyethersulfone, polyvinylidene fluoride, phenol-formaldehyde, urea-formaldehyde, melamine-formaldehyde, epoxy and unsaturated polyester resins are used in the construction of electric components, while other plastics like polypropylene (PP), low-density polyethylene (PE), polyvinyl chloride (PVC), polyurethane (PU), ethylene propylene diene monomer (EPDM), silicone, etc. are applied as jacket materials for electrical cables or for the construction of pneumatic hoses and compressed air tubing. PP, PE and PVC are also used to construct drain pipes, while shields of polycarbonate can protect the food area below light sources from shattered glass after accidental breakage of lamps. Silicone, nitrile, PU, EPDM and butyl rubber are largely used as materials for gaskets, seals, etc. Epoxy is widely used as floor, wall and ceiling coatings. Remember that many plastics perform differently at -25 °C than they do at 20 °C.

# Integration of Piping

Utility piping in technical corridors or zone H areas should be integrated into wall compartments or the ceiling. If this is not possible, it is recommended to use open racks, fixed to the ceiling, or walls and columns close to the ceiling. However, sufficient clearance must be provided between pipe runs and adjacent surfaces so that both are readily accessible for cleaning and maintenance. The pipe racks must be designed hygienically to minimize the presence of horizontal ledges, crevices or gaps where inaccessible dirt can accumulate.

Food processing support piping should be directly routed from service rooms to process areas and should always be logical and simple. The amount of utility piping should be minimized and should have—like process piping—a slope of 1/200 to 1/100. Especially in process, hot water and process steam piping, standing "pools" of liquid that can support the growth of microorganisms must be avoided. To remove condensate, steam traps should be located at all low, convenient points along any extended pipe length. Steam purges for relief of steam condensate in a drain should be closely connected to that drain. In open systems, the steam vapor coming out of a drain can cause humidity and odor problems within the factory. Discharge of condensate from the system should be via an air break to prevent back-siphonage. Neither process nor utility piping should have dead legs.

Like process piping, utility piping should be grouped together in easily accessible pipe trains whenever possible. The points of use should also be grouped, in an attempt to minimize individual ceiling drops. Vertical entrance of piping into the equipment or equipment jacket is more hygienic than horizontal utility piping runs. Running of process and utility piping over open equipment in food preparation areas cannot be accepted, and nesting of ductwork should be avoided. Piping should not clutter the ceiling. When necessary, suspended racks that run over a product zone shall be equipped with a drip pan that protects the product zone below and can be readily removed for cleaning. Bumper guard construction can also be installed in heavy traffic areas to protect piping from external mechanical forces.

Piping should be installed at least 6 cm from walls and floors to encourage thorough cleaning around it. Piping in corners should be avoided, as it hampers thorough cleaning. Process equipment shall be installed such that enough space is provided to facilitate pipe cleaning.

As piping (utility and process) can affect or disrupt the airflow pattern in zone H rooms, a fog test can control airflow patterns. The geometry of the utility piping can destroy the desired air pattern (e.g., piping with a square or rectangular profile is less favorable than circular). Square and rectangular shapes create turbulence and depressions where dust can accumulate, but cylindrical profiles make cleaning easier.

## Penetration of Piping through Walls, Ceilings and Floors

Piping that transports dirty fluids should not run in the vicinity of or cross utilities that transport process aids, especially if these process aids are in direct contact with the food to be processed. Like process piping, food processing support piping should run unidirectionally, with the support piping running from the cleanest area toward the least clean areas. Support systems should deliver a certain process aid first in the process area with the highest hygienic risk (zone H) and last in the zone of lowest hygienic risk (zone L).

Pipeline penetration through walls, ceilings and floors should be minimized, as holes in these areas can lead to sanitation problems and can invite the entry of insects and rodents. Openings in floors for pipes should be guarded with a sleeve to avoid spill of cleaning solutions onto a lower floor. When several pipes penetrate the floor, a larger curbed floor can replace several pipe sleeves to improve the cleanability of the surrounding process environment. However, that curbed floor may create a large opening where pests may harbor, and where dirt, water, etc. may accumulate. It must be a completely closed curb with a cover that leaves no gap around the penetrating piping.

Holes in walls for pipe traverse need not to be sealed water- and air-tight when both sides of the wall are in rooms of the same hygienic zoning, but any opening should be large enough for access and cleaning. However, if a wall separates rooms of different hygienic zoning, all holes for pipe traverse must be sealed. The exterior surfaces of the pipes that traverse walls or ceilings should then have water- and air-tight contact with the wall or ceiling. Foaming-in-place is an appropriate method to close the gaps formed between pipe surfaces and walls as are the applications of plastic caps around the piping and flashing flanges. If running of process and utility piping through walls or ceilings in zone H rooms cannot be avoided, the apertures through the walls and ceilings shall be properly closed against air leakage, as they give excessive air volume losses which may affect product.

## Sanitary Insulation of Piping

Hot piping should not run in the neighborhood of piping that transports cold food products, cold process water, etc. The warm-up of these cold liquids can give rise to the growth of food pathogens. Insulation of hot piping is required, not only to economize on energy, but also to prevent excessive heating of the food production environment above acceptable temperatures. Poorly insulated ethylene glycol and cold/chilled water piping can sweat or be covered with ice, resulting in dripping water. To avoid ingress of dust, vermin, etc. into the insulation, it is highly recommended fully welded metal cladding or plastic covering be installed. It should be impossible to walk on the insulation during maintenance. Damage to insulation can be inhibited by covering the pipe insulation with a smooth, hard, non-electrostatic, plastic cover, rather than steel sheet cladding.

## Hygienically Designed Transfer Panels

Flexible hoses can be used for performing transfers within a given process area. However, hoses are impractical to perform transfers between rooms, especially if these rooms have a different level of "cleanliness." To make connections between different processing units in adjacent rooms, the use of hygienically designed transfer panels is recommended. Interconnection between the different ports should be made with sanitary U- and J-bends. Piping behind the transfer panel and the panel ports must be sloped to ensure proper drainage of residual liquid toward a drain pan. For the same reason, the whole transfer panel can tip a little bit forward. Ports should be capped when not in use to prevent any potential spill or contamination.

## Chemical and Wear-resistant Floors

Floors should be sloped toward drains and provided with curbed wall floor junctions, with the curbs having a 30-degree slope to prevent accumulation of water, dust or soil.

Concrete flooring, including the high-strength granolithic concrete finishes, are especially suitable in warehouses where excellent resistance to heavy traffic is critical. However, untreated concrete can be dusty if dry and highly susceptible to damage from water and acids when wet. Concrete flooring is not recommended for high-care production areas, because it can spall and absorb water and nutrients, allowing microbial growth below the surface.

Epoxy flooring provides a durable, seamless, chemical-resistant and readily cleanable surface. However, over time the coating can crack and buckle due to exposure to cleaning chemicals or wear caused by heavy traffic. Once this happens, moisture pockets under the coating can create a microbiological niche.

Tile flooring is an excellent surface for food plants. However, with heavy wear and in more aggressive cleaning environments, tiles may lose some of their grouting, allowing the penetration of water beneath them. Plastic or asphalt membranes may be laid between the underlying concrete surface and the tiles. Brick floors also may be satisfactory but tend to be somewhat fragile and, unless vitrified, permit water penetration.

Welded PVC sheets have excellent chemical resistance. However, they are not suitable in hot and wet areas, and the welded PVC may be damaged by heavy cart traffic. Steel plates may be used on balconies, for example, and on loading docks and walkways in the vicinity of the process. However, they may corrode and are difficult to bond to concrete. Wood floors are satisfactory in packing and warehouse areas; however, the wood should be impregnated and coated with a durable plastic such as PU. Generally, wood floors may become worn, porous and absorbent, requiring expensive maintenance, and thus are not typically installed in modern food plants.

## Pocket-free Drains

Drains should have appropriate capacity to avoid "ponding" of water and hence contamination in the area to be drained. The drain bodies must be free of pockets that can hold food soil; otherwise, they will cause odor problems. Only drains with an internal P-trap and atmospheric break should be used. P-traps create a water-lock that keeps sewer gases out of the plant.

## Balanced Air Supply and Exhaust System

Exhaust systems should have sufficient capacity to remove excess heat, dust, vapor, aerosols, odors and bioburden from process rooms. However, a positive overpressure must always be maintained. The supply of filtered air in the room by the heating-ventilation-air conditioning system must thus be large enough, otherwise the exhaust system will attempt to draw the required amount of air from adjacent less clean areas through doorways and windows. Exhaust fans must be located outside the building to maintain a negative pressure in the portion of the duct system located within the building. If they are installed in the exhaust hood, the exhaust air is pushed through the duct and not pulled out. By pushing vapors, fumes, etc. through that duct, the system puts the exhaust duct under positive pressure, which can force dirty air back into the room through holes and gaps in the duct work.

## Hygienically Designed Lighting

Lighting must illuminate horizontal and vertical working surfaces evenly, without causing glare and at an intensity of about 300–500 lux at normal working height. Walls and ceilings should be light-colored because that permits fast detection of dirt and soil on their surfaces. In contrast, dark-colored walls and floors require additional lighting.

Preference should be given to lighting mounted on ceilings rather than on walls, because process equipment, storage racks, etc. can form shadows that make cleaning and inspection of floor, walls or ceilings difficult. For the same reason, overhead piping may not obstruct lighting.

Selected lighting should produce little heat and UV light to prevent attraction of insects. Because high-intensity discharge lamps (metal halide, and high- and low- pressure sodium lamps) have high penetration depth, they are used as high-bay lighting in warehouses; fluorescent luminaires are preferred as low-bay lighting, giving good illumination with less glare when covered with a prismatic cover or opalescent diffusing panel.

Lighting systems and their supports may not create horizontal ledges, legs or surfaces. To avoid projections that can accumulate dust, they can be built into the ceiling or wall with a hermetically closed seal, a procedure that is typical for cleanroom areas where lamps are changed via the technical area.

## Hygienic Supply and Application of Electricity

In zone M areas, installing individual cables or multiple cables of small diameter, sharing the same route, in conduits is recommended. When two or more cables partly share a common route but go to different termination points, the creation of unsealable openings that allow the cables to enter or exit the conduit is possible. However, this practice is only recommended for short distances. For long distances, straight line, non-bundled electric cables should be mounted on wire trays, preferably separated from each other. Vertical cable trays are less prone to dust accumulation, and are more accessible for inspection and cleaning. The use of horizontal racks for electrical cabling should be minimized, or they should be protected by a removable lid or installed vertically (on their side) to minimize horizontal surfaces.

When two or more cables partly share a common route, but go to different termination points, unsealable openings allowing cables to enter or exit the conduit should be avoided. Conduits should be suitably sealed at both ends with a proprietary cable/sealing gland where a cable does pass through. In the food contact and splash areas, cables can also be protected from dirt, penetrating liquid and damage by encapsulating them in hermetically closed cable housings. However, the use of pipe rather than conduit should be discouraged because of the difficulties in maintaining the integrity of the piping system at cable entries and exits. Cable mounting in pipes still creates a hollow body and hence a hygienic risk.

Electric components should be enclosed in dust- and water-tight cabinets or field boxes with all connections made at the bottom. Connections of cables and wires to housings must be sealed. The enclosures should be spaced away from equipment or walls and should be provided with an easily drainable 30° top roof. The heat generated by the electrical installations within these enclosures, and concomitantly the dust that penetrates the electrical installation during its cooling by means of fans, should be ventilated toward a technical area or a central ventilation system.

## Control Panels

Control panels with high ingress protection rating should be provided with hygienically designed control and indicator devices. However, the more modern and hygienic membrane panels or touch-screen display panels now often replace these older, non-computer-based control panels.

# Hygienic Design of Equipments

Problems caused by microbial contamination of foods tend to be expensive, particularly if these result in consumer recalls. As a result of the development and application of increasingly mild preservation technologies, processed foods become more sensitive to microbial (re)contamination, requiring greater control of the manufacturing process. One way to achieve this added control is to "build in" hygiene into the equipment used in the food manufacturing facility from the start.

The hygienic design of equipment plays an important role in controlling the microbiological safety and quality of the products made. A hygienic factory should prevent products from having high microbial counts, from containing toxins of microbial origin, and from containing residues of chemicals used for cleaning and disinfection. In addition, the hygienic facility should prevent food from being contaminated with other non-food substances, such as lubricants, coolants and antimicrobial barrier fluids, as well as from containing foreign bodies, such as pieces of metal, plastic, packing material and insects or other vermin, or parts thereof. This may appear a complex task, but with increased activity by international standard-setting organizations more specific and workable information on this topic is now available to the food industry.

## Standards and Guidelines

There is an increase in the involvement of regulatory and advisory bodies in the area of hygienic processing and hygienic design. In the European Union (EU), specific legislation, such as the Machinery Directive 98/37/EC and the Council Directive 93/43/EEC on the hygiene of foodstuffs, requires all food handling to be performed under hygienic conditions. In the framework of the machinery directive, many technical committees (TC) continue many efforts with regard to the preparation of standards.

For example, the TC 153 has prepared the general hygiene standard EN 1672-2: Food Processing Machinery, Part 2: Hygiene Requirements. This European standard sets common requirements in respect of risks to hygiene arising from the use of the machine and process. It primarily covers general aspects of hazards to the food created by the machine in order not to introduce hazards to the consumer of the food. Other specific requirements currently being prepared or already published include standards pertaining to food processing equipment, such as planetary mixers; food processors and blenders; beam mixers; chop cutting machines; cube cutting machines; and dough mixers.

In addition, the International Organization for Standardization (ISO) TC 199 has prepared a standard titled, "Hygiene Requirements for the Design of Machinery." This international standard specifies hygiene requirements of machines and provides information for the intended use to be provided by the manufacturer. It applies to all types of machines and associated equipment used in applications where hygiene risks to the consumer of the product can occur. However, it does not cover requirements relative to the uncontrolled egress of microbiological agents from the machine.

The European Hygienic Engineering and Design Group (EHEDG) has developed similar design criteria and guidelines on the hygienic design of equipment and hygienic processing. The EHEDG has links with the Comité Européen de Normalisation (CEN), 3-A Sanitary Standards, NSF International and ISO. As these international standard-setting organizations continue efforts to specify hygienic design requirements, food processors will be able to more effectively select and introduce such equipment into their facilities.

## Selected Hygienic Design Criteria and Requirements

There are several aspects of designing hygiene into equipment that should be considered by the food processor before reengineering or introducing process equipment into the plant. In general, construction materials that may come in contact with food should not be able to make a food product toxic. Equipment must be designed to be self-drainable to make it possible to remove all residues of products and chemicals. To be cleaned without difficulty, surfaces must be smooth and free from crevices, sharp corners, protrusions, and shadow zones. When surfaces are not clean, microorganisms may be protected from destruction by heat or chemicals.

Selected criteria and basic requirements for a variety of hygienic equipment characteristics provide a fundamental overview of areas that can be addressed by food manufacturers:

Materials of construction: Materials used for the construction of a food processing plant must fulfill certain specific requirements. Product-contact materials must be inert to the product under operating conditions, as well as to detergents and antimicrobial chemicals (sanitizers) under conditions of use. They must be corrosion-resistant, mechanically stable, and such that the original surface finish is unaffected under all conditions of use. In addition, non-contact materials shall be mechanically stable, smoothly finished and easily cleaned. The reinforcement in plastics and elastomers should not be allowed to contact the food product.

Surface roughness: Product contact surfaces should be smooth enough to be easily cleanable. The roughness (or smoothness) of a surface usually is expressed in μm, as Ra-value. Generally, the cleaning time required increases with surface roughness. The American 3-A Sanitary organization and the EHEDG specify that food contact surfaces have a maximum roughness of Ra = 0.8 μm. To achieve this quality of surface, polishing or other surface treatment may be required. Cold-rolled stainless steel sheet material, used for vessels and for piping, usually has an Ra-value between 0.2 and 0.5 μm, and thus, further treatment is not needed. According to the EHEDG, rougher surfaces can be acceptable if tests have shown that the required cleanability is achieved. Porous surfaces usually are unacceptable. To be cleaned without difficulty, surfaces must not only be smooth but also free from crevices, sharp corners, protrusions and shadow zones. This applies not only when equipment is new, but during its entire functional lifetime.

Crevices: Crevices cannot be cleaned, and as such, will retain product residues that may effectively protect microorganisms against inactivation. In some cases, crevices are unavoidable. This may be the case if slide bearings must be in contact with product; for example, as bottom bearings of top-driven stirrers or as bearings in scraped-surface heat exchangers. The presence of slide bearings should be considered when writing procedures for cleaning and disinfection. These procedures may require instructions for both partial or total dismantling of equipment, or for increased cleaning times.

In most cases, crevices are the result of incorrect choices when designing (or selecting) equipment. When parts of equipment must be mounted together, metal-to-metal contacts (other than welds) must be avoided because they leave very narrow and deep crevices. Elastomers should be used between metal components, but not in the form of O-rings in standard O-ring grooves, as this, too, will create crevices. The elastomeric material must be mounted in such a way that the seal is at the product side and excessive compression is prevented to avoid destruction of the elastomer. This can be achieved by including design features that align the surfaces of the various parts and provide a metal stop.

Screw threads: The use of screw threads and bolts in the product area should be avoided. Where unavoidable, the crevices created should be sealed, at minimum.

Sharp corners: Sharp corners in the product area should be avoided. Exceptions are constructions where the sharp corner is continually swept, such as in lobe pumps. Welds should not be made in corners, but on the flat surfaces, and must be smooth.

Dead areas: There is a significantly reduced transfer of energy to the food residues (soil) in dead areas in process equipment that is placed outside of the main flow of cleaning liquids than there is to the soil in the main flow. Such areas are difficult to clean, and therefore, should be avoided. If unavoidable, their presence should be taken into account when devising the cleaning procedures. Typical shadow zones, for example, can be found in the legs of T-pieces in pipelines, which are used to mount sensors such as pressure gauges.

Drainability of equipment and process lines: To make it possible to remove all chemicals from process equipment, the equipment must be designed to be self-drainable. Thus, surfaces and pipes should not be completely horizontal, but slope toward drain points. There should be no ridges that may hamper draining. Where it is not possible to build equipment in such a way that proper draining is possible, procedures must be designed to ensure that residues of cleaning and disinfection liquids can be removed in another way. The method used should be well documented with clear instructions. Draining also is important, even in cases where no chemicals are used, because many microorganisms can easily grow in residual water, needing only minute amounts of nutrients to multiply.

Top rims of equipment: The design of the top rims of product-containing equipment must avoid ledges, where product can lodge and that are difficult to clean. Open-top rim design must be rounded and sloped for draining. If the top rim is welded to the wall, the weld must be flush and polished to provide a smooth surface. In this case, the rim must be totally closed. The weld must be continuous and any holes must be sealed by welding, gaskets or plastic caps.

Mandoor covers: Mandoor covers intended to protect the food products may accumulate dirt, which will enter the product in the vessel when the lid is opened. Policy should specify that no tank is opened during production unless absolutely necessary.

Shaft passages and seals: Shaft passages and seals may leak product to the outside of the line. Microorganisms may then multiply in the product and grow back to the product side. In the case of dynamic seals, such as those for shafts of valves, pumps and mixers, the movements of the shaft will assist the transfer of product to the outside and the transfer of micro-organisms to the product side. This applies to reciprocating shafts, and to a lesser extent, to rotating shafts, the latter always displaying some axial movement. Reciprocating shafts can be sealed by means of flexible diaphragms or bellows. To prevent the ingress of microorganisms in rotating shafts, double seals with microbiocidal barrier liquids should be used. If not replaced in a timely manner, however, such barriers may become a growth medium for microorganisms.

## References

- Food-processing: newworldencyclopedia.org, Retrieved 05 January, 2019

- Hygienic-design-of-equipment-in-food-processing: foodsafetymagazine.com, Retrieved 23 June, 2019

- Boye JI, Arcand Y (2013) Current trends in green technologies in food production and processing. Food Engineering Reviews 5: 1–17

- Hygienic-design-of-equipment-in-food-processing: foodsafetymagazine.com, Retrieved 18 August, 2019

- Luning, P.A., Marcelis, W.J. and Jongen, W.M.F. (2005) Food quality management: a techno-managerial approach. WNT, Warszawa

# Ambient Temperature Processing

Ambient temperature processing is the method of food processing that destroys the microbial growth without any change in the temperature of the surrounding environment. Centrifugation, size reduction, coating of food and mixing of food are some of the concepts that fall within this processing. This chapter has been carefully written to provide an easy understanding of these concepts of ambient temperature processing.

## Food Preparation

People process foods every day when preparing meals to feed their families. However, the term "food processing" is broader than preparing and cooking foods. It involves applying scientific and technological principles to preserve foods by slowing down or stopping the natural processes of decay. It also allows changes to the eating quality of foods to be made in a predictable and controlled way. Food processing uses the creative potential of the processor to change basic raw materials into a range of tasty attractive products that provide interesting variety in the diets of consumers.

All food manufacturers should make safe foods so that consumers are not at risk. This is not only microbiological risks, but also glass splinters, pesticides or other harmful materials that can get into the food and lower its quality. Consumers consider *eating quality* as the main factor when buying foods, and a food should fit in with traditional eating habits and cultural expectations of texture, flavour, taste colour and appearance. For some foods, *nutritional quality* (e.g. protein content, vitamins and minerals, etc.) is an important consideration. Product quality is affected by the raw materials, the processing conditions and the storage and handling that a food is subjected to after processing. Food processors should understand the composition of their foods because it enables them to predict the changes that take place during processing, the expected shelf life of the product and the types of microorganisms that can grow in it. This information is used to prevent food spoilage or food poisoning. Details of the composition of raw materials or products can be obtained from university food science departments, bureaux of standards or food research institutes.

### Types of Processing

Without processing, as much as 50 to 60 percent of fresh food can be lost between harvest and consumption. This may be due to inadequate storage facilities, which allow micro-organisms or pests to spoil the stored food. Improved storage can greatly reduce these losses. Processing methods that are suitable for village scale processing can be grouped into six categories. A number of other preparation methods (such as mixing, coating with batter, grinding, cutting, etc.) alter the eating quality of foods, but do not preserve them. It is important to note that the production of most food processed foods uses more than one of the categories in Table. For example, jam making involves heating, removing water, increasing the levels of acidity and sugar, and packaging. Smoking fish or meat involves heating, removing water and coating the surface with preservative smoke chemicals.

## Effect of Processing on the Quality of Foods

In addition to preserving foods, secondary processing alters their eating quality. A good example is cereal grains, where primary processing by drying and milling produces flour, which remains inedible. Secondary processing is used to produce a wide range of bakery products, snack foods, beers and porridges, each having an attractive flavour, texture and/or colour. Eating quality is the main influence on whether customers buy a product. Foods that have an attractive appearance or colour are more likely to sell well and at a higher price. It is therefore in the interests of processing businesses to find out what it is that consumers like about a product using market assessments and ensure that the products meet their requirements.

## Scales of Operation

When operating as a business, food processing can take place at any scale from a single person upwards. The focus of this booklet is on the smaller scales of operation from "home-scale" to "small-scale".

## Home-scale Processing

Foods that are intended for household consumption are usually processed by individual families or small groups of people working together. Many of the world's multinational food conglomerates started from a single person or family working from home. In developing countries, home-scale processors aim to generate extra income to meet family needs such as clothing or school fees. Where this is successful, many later expand production and develop first into a micro- or small-scale business, and later into larger scale operations.

Characteristically, home-scale processors cannot afford specialized food processing equipment and rely on domestic utensils, such as cooking pans and stoves for their production.

Table: Types of village food processing.

| Category of process | Examples of types of processes |
|---|---|
| Heating to destroy enzymes and micro- organisms. | Boiling, blanching, roasting, grilling, pasteurization, baking, smoking. |
| Removing water from the food. | Drying, concentrating by boiling, filtering, pressing |
| Removing heat from the food. | Cooling, chilling, freezing. |
| Increasing acidity of foods. | Fermentation, adding citric acid or vinegar. |
| Using chemicals to prevent enzyme and microbial activity. | Salting, syruping, smoking, adding chemical preservatives such as sodium metabisulphite or sodium benzoate. |
| Excluding air, light, moisture, micro-organisms and pests. | Packaging. |

Table: Scales of commercial food processing.

| Scale of operation | Characteristics |
|---|---|
| Home- (or household-) scale | No employees, little or no capital investment. |
| Micro- (or cottage-) scale | Less than 5 employees, capital investment less than US$1 000. |
| Small-scale | 5-15 employees, capital investment US$1 000-US$50 000. |
| Medium-scale | 16-50 employees, capital investment US$50 000-US$1 000 000. |
| Large-scale | More than 50 employees, capital investment over US$1 000 000. |

They may work part-time as the need for money arises and use part of the house, or an outbuilding for processing. However, in many situations the lack of dedicated production facilities means that there is a risk of contamination and product quality may be variable. This may reduce the value of the processed foods and the potential family income. A role of extension agents and training programmes is to upgrade facilities and hygiene, to introduce simple quality assurance techniques and improved packaging, to enable products to compete more effectively with those from larger processors.

Where families generate sufficient income from sales, some choose to invest in specialist equipment (such as a bakery oven, or a press for dewatering cassava or making cooking oil). In most cases, such equipment can be made by a competent local carpenter, bricklayer or blacksmith. This allows home-scale businesses to expand and become micro- or smallscale enterprises.

## Micro-scale Processing

Whereas home processors may sell their products to neighbours or in village marketplaces, the move up to micro-scale processing requires additional skills and confidence to compete with other processors and to negotiate with professional buyers, such as retailers or middlemen. Similarly, although the quality of their products may be suitable for rural consumers, it may not be sufficient to compete with products from larger companies in other markets. To successfully expand to micro-scale production, village processors need technical skills to make consistently high quality products, and financial and marketing skills to make the business grow and become successful. They may require assistance to gain these skills and confidence, and short training programmes or technical extension workers can help them to establish improved production methods, quality assurance and selling techniques.

## Small-scale Processing

The expansion to a small-scale processing operation requires additional investment to produce larger amounts of product in a dedicated production room. It is likely to require specialist equipment that is either made by a metal workshop in a nearby town or imported, because most rural blacksmiths do not have the necessary skills, equipment or materials to make such equipment. At this level of production, village processors are likely to be in competition with other small-scale businesses, larger companies and imported products. They need to develop attractive packaging, quality assurance techniques, and the financial and managerial skills needed to run a successful small business.

Table: Origins of some of the world's major food-processing companies.

| Year | Food company |
|------|--------------|
| 1200 | Bock beer was invented in the German town of Einbeck, and is still manufactured there |
| 1383 | Lowenbrau brewery opened in Munich, Bavaria and remains in production there today |
| 1715 | French distiller, Jean Martell, began brandy production in Cognac, France |
| 1725 | The chocolate company, Rowntree, had its beginnings in a grocery store in York, UK |
| 1871 | The first margarine factory was opened by Jan and Anton Jurgens at Oss in Holland |
| 1876 | H J Heinz joined his brother and cousin to produce tomato ketchup and pickles |
| 1877 | Acream separator was invented by Swedish engineer, Carl Laval, whose company is now Alfa-Laval |
| 1884 | Swiss miller, Julius Maggi introduced powdered pea and beet soups, later to become "Maggi" Maggi cubes |
| 1899 | Coca-Cola was bottled under contract for the first time by Benjamin Thomas and Joseph Whitehead in Tennessee, USA, instead of syrup being mixed with carbonated water at the point of sale |

If the level of investment at this scale is too high for individual families, an alternative approach is for a group of people, such as a farmers' group, or a women's group, or a producer co-operative to operate the food processing business together. They invest jointly in the equipment and facilities, and market their products under a single brand name. There are many advantages to this approach including a greater willingness by lenders to make a loan if a group is sharing responsibility for the repayments, new employment opportunities for those without land, discouraging migration to larger towns or cities, and providing greater financial security and an improved standard of living to larger numbers of people.

Small-scale processing.

Many governments promote the development of small-scale food processing enterprises because they:

- Have the potential to create significant levels of employment;

- Increase food security for growing urban populations as well as rural families;

- Produce products that can substitute for imported foods or have export potential, and thus help reduce balance of payments problems and improve the overall prosperity of the country.

## Cleaning Foods

Cleaning in the food industry is not an easy task. However, it is a critical step within food production since it is crucial to maintain and guarantee food safety. Understanding various soil challenges, why we clean and how detergents and disinfectants work is key to ensuring a safe, hygienic manufacturing environment.

Here are the reasons why we clean and disinfect food:

- Prevent Transfer of Products/Ingredients: If a number of products are manufactured on the same machine, it is undesirable to cross-contaminate chemicals or alternate from one product to the next.

- Avoid Microbial Contamination: This can lead to a number of problems––reduced product quality, harm to health or even life threatening circumstances in some cases. Cleaning alone is no guarantee of decontamination, but it is a pre-requisite to disinfection.

- Ensure Disinfectant Efficiency: Soil impacts the effectiveness of a disinfectant. The less soil on the surface, the more effective the disinfect will be at reducing microbiological contamination.

- Improve Plant Efficiency: Soil contamination reduces the efficiency of equipment and the production process.

- Increase Safety: Facilities that are not cleaned effectively have more potential safety risks—like slips and falls—due to food waste on floors. Also, major incidents due to build up of soil in equipment can also occur.

- Impact Financial Implications: Reducing waste from spoilage can significantly extend the life of equipment and machinery.

- Minimize Legal Ramifications: Although it may not be common knowledge, there are often legal requirements for food facilities to clean surfaces and equipment to a specific standard.

- Boost Stakeholder Confidence: Finally the appearance of plant and premises is often overlooked but the psychological benefits and confidence gained from clean, hygienic equipment and tidy surroundings have a significant impact on both worker satisfaction and customer confidence.

Cleaning and disinfection should be considered as two discrete steps in the cleaning procedure. Cleaning is the complete removal of residues and soil from surfaces, leaving them visually clean so that subsequent disinfection will be effective. Without effective cleaning, disinfection will be compromised.

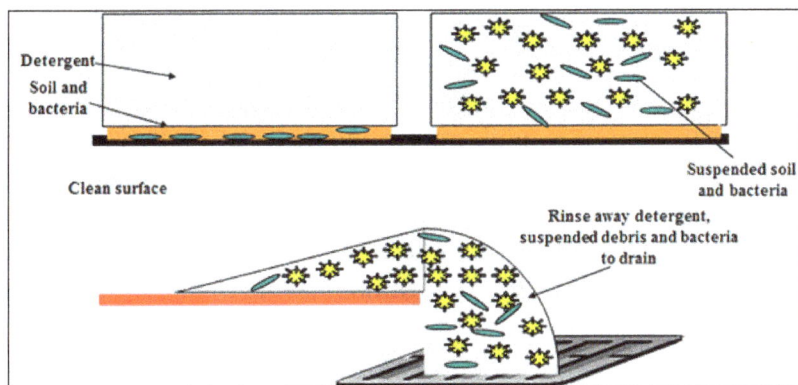

Detergents are used to remove soil from a surface. The soil—a mixture of food waste and bacteria—is on or attached to the surface of the processing equipment, floors or walls. The action of the detergent solution is to suspend this soil and bacteria mixture away from the surface and allow for it to be rinsed off to the drain. However, there are many soils found in the food industry and the cleaning procedure and detergent used in order to achieve the desired detergent action is different for each soil.

The most common soils—carbohydrates like sugar, starch and cellulose—are the easiest to remove. Proteins—meat, milk and eggs—are probably the most difficult because changes in heat and pH alter the structure of the protein and bind it to other molecules, increasing their tenacity and often rendering them insoluble. For example, while milk is soluble in water, if you over boil a pan of milk, the resulting milk soil becomes difficult to remove form the pan.

Fatty soils are not water-soluble and pose a greater challenge than carbohydrates. Here, it's necessary to use alkaline cleaners and elevated temperatures above the melting point of the fat to achieve an efficient clean. Mineral salts––the inorganic food soils––lead to scale formation on equipment. Acidic cleaners are required to efficiently remove the scale.

There are four variables within the cleaning process that can impact its efficiency to remove soil:

1. Detergent/Concentration

2. Time

3. Temperature

4. Physical Action

Devised in 1959 by Dr. Herbert Sinner, Sinner's Circle is universally known as the model to demonstrate that reducing one of the four factors can be compensated by increasing another. For example, you may be able to increase the temperature to enable you to use a lower concentration of chemical.

## Disinfection

Disinfection is the process by which microorganisms are killed so that their numbers are reduced to a level which is neither harmful to health nor to the quality of perishable goods. Following cleaning, surfaces will be free from soil but microorganisms remain. Using validated disinfectants on surfaces, following the instructions and contact times, reduces microorganism levels to the required level for food production.

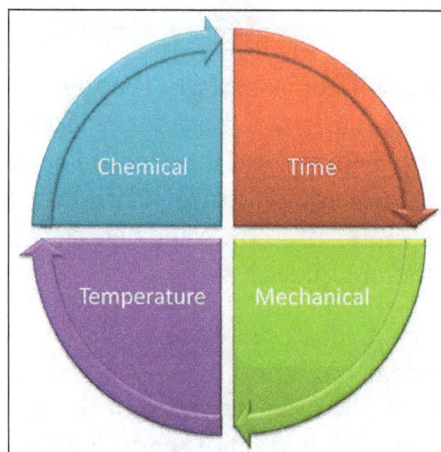

The method by which disinfectants kill the microorganism––referred to as their "mode of action"––varies with the active ingredient. The table on the right shows some of the key activities and their mode.

When selecting a disinfectant, a number of considerations need to be made.

Above all, the approvals each disinfectant has should be taken into account during selection, for example if you are using a product in a chilled environment, the disinfectant should be proven to work at the temperature you are intending to apply it at.

## Cleaning Validation

The validation of cleaning and assuring standard, consistent results has become a topic of high priority in the food processing industry. This validation is used to show proof that the cleaning system consistently will perform as expected. A surface is chemically clean if there are no microscopic residues of soil remaining and no residual detergents or disinfectant chemicals to contaminate the food product. Determination of chemical cleanliness requires tools other than the human eye. Techniques for determining such small amounts of soil include:

- Visual moisture on surfaces: Soiled surfaces become hydrophobic, so water will bead up and show the presence of soil.

- Dyes: Specific dyes with an affinity for certain soils like protein or starch can also be applied to a surface to make the soil visible.

- Optical methods: These are used quite regularly in some food processing operations for detecting soils. Interruption of a light beam, reflectance and absorbance are all optical means of detecting soil.

- Adenosine triphosphate (ATP): ATP levels are present in all organic material. Measuring its presence is a very effective way of providing immediate feedback to ascertain the cleanliness of critical control areas.

The basic principles discussed here should provide insight into the variables that should be taken into account when devising a cleaning schedule. Controlling these variables will present opportunities to deliver real improvements in operational efficiency in food processing, while improving overall food safety.

## Sorting and Grading

Sorting and/or screening (dry and wet) are/is the separation of raw materials and/or food slurries into categories on the basis of shape, size, weight, image and colour. The size sorting and dry cleaning of agricultural raw materials separates solids into two or more fractions on the basis of different sizes, usually by sieving or screening. Size sorting is especially important for food products which have to be heated or cooled, as large differences in size might lead to an over-processing or under-processing of the product.

Sorting also allows the separation at first sight of some (undesirable) additional material (e.g.leaves, stones) or inappropriate raw material (immature or rotted berries), and aims at ensuring that only good quality fruit is preserved and passed through for further processing.

For size sorting, various types of screens and sieves, with fixed or variable apertures, can be used. The screens may be stationary, rotating or vibrating. Shape sorting can be accomplished manually or mechanically with, for example, a belt- or roller-sorter. Weight sorting is a very accurate method and is therefore used for more valuable foods (cut meats, eggs, tropical fruits, certain vegetables). Image processing is used to sort foods on the basis of length, diameter, and appearance, i.e. surface defects and orientation of food on a conveyor. Colour sorting can be applied at high rates using microprocessor controlled colour sorters.

## Grading

Grading is the assessment of a number of characteristics of a food to obtain an indication of its overall quality. Grading is normally carried out by trained operators. Meats, for example, are examined by inspectors for disease, fat distribution, carcasse size and shape. Other graded foods include cheese and tea. In some cases the grading of food is based on laboratory analyses results. In the wine industry, grading also covers the necessary classification of the grapes harvested according to their degree of maturity (for example, sugar content). Many characteristics cannot be examined automatically and trained operators are employed to simultaneously assess several characteristics in order to produce a uniform high-quality product. Grading is more expensive than sorting (which looks at only one characteristic) due to the high costs of the skilled personnel required.

## Peeling

The objective of peeling is to remove unwanted or inedible material from vegetable raw materials. This improves the appearance and taste of the final product. During peeling, peeling losses need to be minimised by removing as little of the underlying food as possible but still achieving a clean peeled surface.

## Field of Application

Peeling is applied on an industrial scale to fruits, vegetables, roots, tubers and potatoes.

Various methods for peeling exist: flash steam peeling, knife peeling, abrasion peeling, caustic peeling and flame peeling.

## Steam Peeling

Steam peeling is carried out either as a batch-wise or a continuous process. The raw materials (roots, tubers) are treated in a pressure vessel and exposed to high-pressure steam (180°C to 200°C). In the case of tomatoes, the temperature may be lower (120 -130°C). The high temperature causes a rapid heating and cooking of the surface layer (within 15 to 30s). The pressure is then instantly released, which causes flashing-off of the cooked skin. The continuous steam peeler is a pipe with a screw inside. The steam is feed direct into the pipe (generally at lower pressure than the batch process) and the product is heated during the residence time (adjustable). Most of the peeled material is discharged with the steam. Any remaining traces are sprayed off with water.

## Knife Peeling

The materials to be peeled (fruits or vegetables) are placed onto a rotating disc and pressed against stationary or rotating blades to remove the skin. Knife peeling is mostly used for citrus fruits, as the citrus skin is easily removed and the fruit suffers little damage.

## Abrasion Peeling

The material to be peeled is fed onto abrasive rollers or fed into a rotating bowl which is lined with an abrasive. The abrasive surface removes the skin, which is then washed away with water. The process is normally carried out at ambient temperature.

## Caustic Peeling

The material to be peeled is passed through a dilute solution (1 to 2%) of sodium hydroxide. This treatment softens the skin, which can then be removed by high-pressure water sprays. A new development in caustic peeling is dry caustic peeling. The material is dipped in a 10 % sodium hydroxide solution. The softened skin is then removed by rubber discs or rollers. A drawback of caustic peeling is that it causes decolourisation of the product.

## Flame Peeling

A flame peeler utilises a conveyer belt to transport and rotate the material through a furnace heated to temperatures above 1000°C. The skin (e.g. paper shell, root hairs) is burned off and then removed by high-pressure water sprays. Flame peeling is used, for example, for peeling onions.

# Extraction of Food Components

Extraction is a process that is growing in importance. It is generally more energy efficient than competitive processes such as expression—the pressing of biological feed materials to liberate fluids. For example, sugar is extracted from sugar beets with hot water, which yields a sucrose stream free of contaminants and of higher concentration (typically 15% sugar) than can be achieved by expression. Solvent extraction can be made selective for specific components of the feed. For instance, supercritical carbon dioxide (SC-CO$_2$) will selectively dissolve caffeine from coffee beans to yield decaffeinated coffee. The extracted caffeine can then be recovered for sale as a pharmaceutical. Extraction can recover thermally labile components that would be degraded by heating, such as gelatin from collagen. Table gives some examples of typical extraction processes employed industrially.

Table: Some examples of industrial extraction processes.

| Solvent | Feed | Product | Component |
|---|---|---|---|
| Water | Apple pulp | Apple juice | - |
| | Malted barley | Brewing worts | Sugars, grain solutes |
| | Kelp | Carrageenan | - |
| | Manioc | Cassava | Cyanogenetic glycosides |
| | Citrus press residues | Citrus molasses | - |
| | Papaya latex | Papain | Papain |
| | Rosemary leaves | Rosemary essential oil | Rosemary essential oil |
| | Citrus peel | Citrus essential oils | Citrus essential oils |
| Acidic water | Collagen | Gelatin | Gelatin |
| | Citrus peel | Pectin | Pectin |
| | Hog stomach | Pepsin | Pepsin |
| Alkaline water | Defatted soy flour | Soya protein | - |
| Aqueous ethanol | Red beets | Betalains | Betalains |
| | Animal pancreas | Insulin | Insulin |
| | Spices | Spice extracts | - |
| | Vanilla beans | Vanilla essence | - |
| Methylene chloride | Green coffee beans | Decaffeinated coffee | Caffeine |
| Supercritical CO$_2$ | Green coffee beans | Decaffeinated coffee | Caffeine |
| | Hops | Hops extract (resin) | Hops essential oils (myrcene, humulene, caryophyllene, and farnesene), alpha and beta acids |
| | Ginger rhizomes | Ginger extract | Gingerols |
| | Pomegranate seeds | Pomegranate seed oil | Pomegranate seed oil |
| | Vanilla beans | Vanilla essence | |

| | Spices (turmeric, nut-meg, mace, cardamom, etc.) | Spice extracts | |
|---|---|---|---|
| | Egg yolk | Decholesterolized egg yolks | Cholesterol |
| | Wheat germ | Wheat germ oils rich in tocopherols | - |
| Hexane | Soybeans | Soybean oil | - |
| Methyl ethyl ketone | Spices | Spice oleoresins | - |
| Tributyl phosphate | Phosphoric acid | Food-grade phosphoric acid | - |

## Physical Principles of Extraction

## Nomenclature of Extraction

A component that it is desired to be removed from the feed through extraction is called the "solute." The phase that is mixed with the feed to remove the solute is the "solvent." After the solvent has been mixed with the feed and the solute has transferred from the feed phase into the solvent phase, the solvent phase is called the "extract" and the feed phase is now called the "raffinate." It must be stressed that, in food processing, the feed is usually solid, semisolid, or gel-like, whereas much of the science of extraction is based on liquid feeds. However, there are very close parallels provided allowance is made for the impact of the nature of the feed on mass transfer properties. Indeed, much of the rest of this text is concerned with means of improving the rate of mass transfer so that the science derived from liquid feeds may be better applied to the processing of food products.

## Solubility

When a feed containing a solute is contacted with a solvent in which the solute is reasonably soluble, then the solute will distribute itself between the feed and the solvent until there is equilibrium between the feed and the solvent phases. When this occurs, the chemical potential of the solute in each phase is the same. The chemical potential is made up of two terms—the concentration of the solute and its activity in the phase concerned. However, in processing foods, it is rarely possible to measure the activity of the solute in the feed; thus, the primary concern is with the solubility of the solute in the solvent. Figure, for instance, shows the solubility of caffeine in SC-$CO_2$ and in SC-$CO_2$–ethanol mixtures. The solubility increases with pressure and with the addition of ethanol to the solvent, but decreases with temperature.

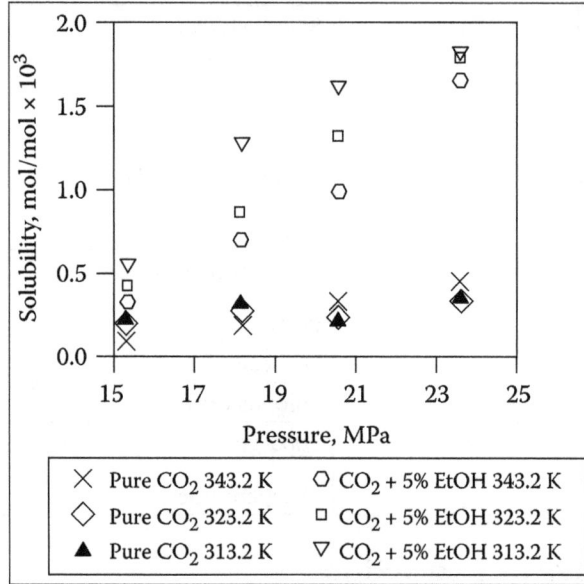

Solubility of caffeine in SC-CO$_2$ and CO$_2$–ethanol.

Determination of solubility is covered in many standard texts. Today, there are powerful packages for estimating chemical properties, which have a wide range of experimental data and strong theoretical bases to permit almost any solubility to be estimated.

It is, however, useful to understand how solubility can be represented graphically. Figure shows the triangular coordinates used to represent solubility. This is an equilateral triangle, with the concentrations of each component along each side as shown.

The composition at point M is then 50% solute, 20% solvent, and 30% feed. A useful feature of such diagrams is that mixtures can be represented by straight lines, and the result of mixing two streams of different composition is determined by simple geometry. Consider two streams of masses D and E and with compositions given by points D and E. Then the result of mixing these two streams will give a composition that lies on the straight line DE and has a composition given by point F such that E/D = DF/FE. This feature will be used in much of the discussion that follows.

The equilibrium between the various phases can readily be shown on triangular coordinates. Figure illustrates this.

The line WXY represents the boundary between a single-phase and a two-phase region. Above the curve, the presence of the solute allows the feed and solvent phases to dissolve in each other, which is of course undesirable from the point of view of separation. The difference between the single-phase and the two-phase region is critically important in extraction because it is the existence of a second phase that permits extraction to take place. The challenge in selecting a solvent is often to find one that will give as large a two-phase region as possible. Ideally, the feed and solvent will be essentially immiscible, and the presence of the solute will not change this immiscibility. However, in many biological systems, it is difficult to reach this ideal state. It then becomes necessary to allow for the effect of a single-phase region and operate so that its influence is minimized.

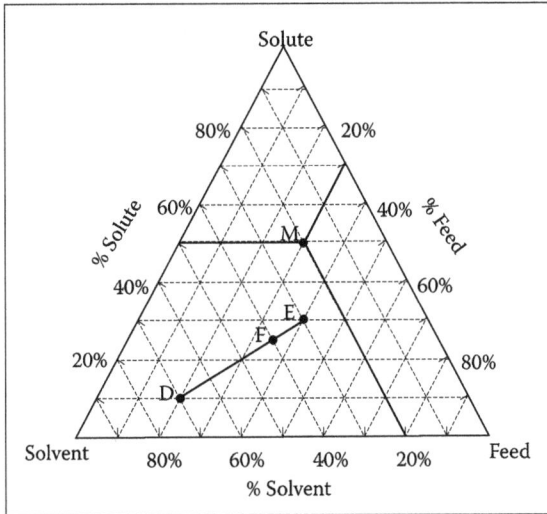

Use of triangular coordinates.

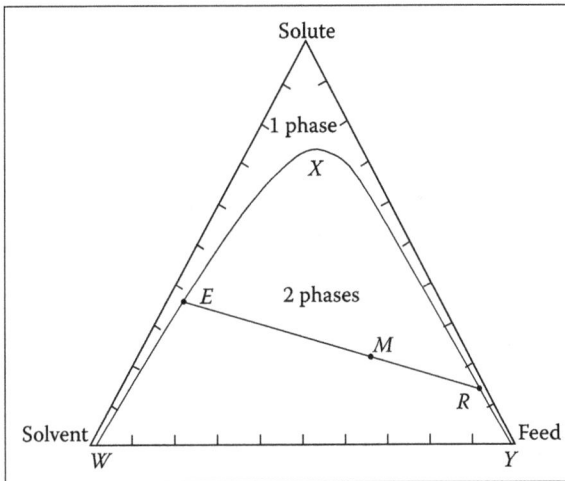

Solubility representation in a ternary diagram.

Because the side of the triangle between the solvent and feed corners represents the line of zero solute, then point W represents the solubility of the feed in the solvent and point Y represents the solubility of the solvent in the feed. Line WXY represents the boundary of the two-phase region.

Consider mixing some solvent with a feed, such that the average composition of the mixture is given by point M in the diagram. Then, provided M is within the two-phase region, the mixture will split and give two phases whose compositions will be given by points E and R on line WXY. These represent the concentrations in the extract and raffinate, respectively. Line ER is called a "tie line," and it joins the composition of two phases at equilibrium. There is a whole series of such tie lines, depending on the starting conditions. EMR is a straight line, as indicated previously, and the mass ratio of solvent to feed is given by the ratio of the lengths EM/MR.

These diagrams refer to conditions at a constant temperature, and the equilibrium line WXY is often referred to as an "isotherm" for this reason.

## Mass Transfer

During the process of extraction, one or more compounds ("solutes") transfer from the biological feed material into the solvent. The physical process underlying the transfer is that the concentration* of the solute in the solvent is less than its concentration in the feed, so that the solute diffuses from the feed into the solvent. However, the diffusion process is hindered by a number of phenomena.

First, there will be some interface between the feed and the solvent. The feed may also be liquid, but the interface between two liquids will slow the diffusion. Consider two liquid phases fully mixed in the bulk of the phases. Thus, the concentration in the bulk of each phase will be the same everywhere in that phase, except close to the interface where diffusion is occurring.

On the feed side, there is a concentration difference between the bulk concentration and the concentration at the feed side of the interface, and it is in this non–fully mixed zone where diffusion takes place. The width of the diffusion zone may be less than a millimeter. Similarly, on the solvent side, there is a concentration difference between the bulk concentration and the concentration at the solvent side of the interface, and diffusion occurs across this narrow layer. At the interface itself, there is a drop in concentration that reflects the difference in chemical potential on either side of the interface.

Thus, three physical layers resist the transfer of mass from the feed to the solvent:

- The diffusion layer on the feed side.

- The resistance to transfer of the interface itself.

- The diffusion layer on the solvent side.

In the case of food processing, the last of these is of little significance. The solvent is generally a liquid of relatively low viscosity, which means that it can relatively simply be fully mixed and the thickness of the boundary diffusion layer can be reduced to a minimum. The challenge in applying extraction to food processing is to minimize the first two of the resistances—that in the feed phase and that at the interface.

In food processing, the feed phase is generally not a low-viscosity fluid. It may be semi-fluid or gel-like. It may be semifluid contained within cellular structures. It may be quite solid. Whatever its state, it will resist mass transfer to a far greater extent than a low-viscosity fluid.

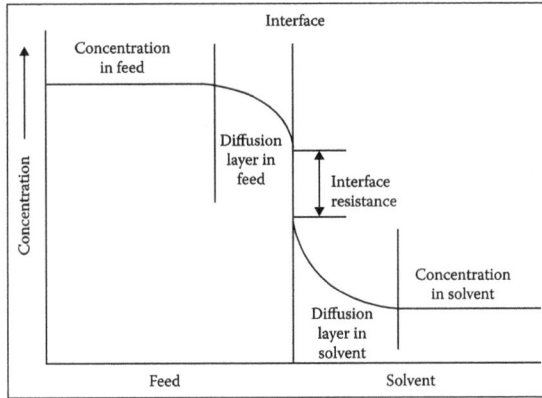

Transfer of solute between two liquid phases.

Similarly, in food processing, the interface is generally not the simple interface between two low-viscosity fluids. It may, as we have seen above, be the interface between a solvent and a semisolid or even a solid. It may be attractive to surfaceactive substances present in the feed, which thicken the interface and thus increase its resistance. It may be a cell wall designed by nature specifically to resist the release of the desired solute. It may even be a solid and even, in some cases, a crystalline solid.

The reason for being concerned about mass transfer is that the slower the rate of mass transfer, the longer the feed and solvent must be in contact. This means, other things being equal, that the longer the two are in contact, the larger must be the equipment in which the contact takes place—and larger equipment is inherently more expensive than smaller equipment. Also, the longer the two are in contact, the greater the chance of the solvent dissolving other solutes from the feed, or for the desired solute to be degraded by temperature or exposure to the atmosphere—or even by reaction with the solvent.

## Diffusion

The rate of diffusion of a single species in a single fluid can be described by Fick's law.

$$J_A = -cD_{AB}\frac{dx_A}{dz}$$

Where $J_A$ is the rate of diffusion (mol m$^{-2}$ s$^{-1}$), c is the concentration (mol m$^{-3}$), DAB is the diffusion coefficient of solute A in solvent B (m$^2$ s$^{-1}$), $x_A$ is the mole fraction of A in B, and z is the direction of diffusion.

What this makes clear is that the greater the area over which diffusion can take place, then the greater will be the rate of mass transfer. Thus, when extracting from a liquid feed, the solvent and the feed are agitated together to increase the surface area between the two to maximize the rate of transfer. Similarly, when extracting solute from a solid feed, the solid should be reduced in size as far as possible to maximize the area through which mass transfer can take.

There are, of course, some constraints on this maximization of surface area to enhance the rate of extraction. With two liquids, for instance, it is possible to mix them so intimately that one emulsifies in the other, so that separation of the solvent after extraction becomes difficult. With a solid feed, the very process of size reduction may be so energy intensive that labile substances are altered. While as large an area as possible can enhance the rate of extraction of a desired solute, it may also make the rate of extraction of a less-desired solute sufficiently high that the desired solute is unacceptably contaminated. Nevertheless, the general rule holds good—a large surface area will speed the rate of extraction.

## Choice of Solvent

The solvent should, naturally, be capable of dissolving the desired solute. It is useful if the solubility of the solute in the solvent is high because this will reduce the quantity of solvent needed to extract a given quantity of solute. However, this is not an essential requirement. Other factors guide the choice of solvent. For instance, in many cases, it is desirable to ensure maximum selectivity—that is, that the solvent dissolves the desired solute preferentially to other potentially soluble materials present in the feed. Water is generally non-selective, but in some cases, as Table illustrates, even it can be sufficiently selective given the right feed and solute.

An important requirement is that the solvent should be reasonably stable and that it should not react chemically with the solute in such a way as to adversely affect the properties of the solute. There are classes of solvents that are inherently acidic or basic, and they can be used to extract anionic or cationic solutes, respectively. Contact with an acidic or basic solution will then recover the solute and regenerate an acidic or basic solvent. In these cases, there is a chemical reaction between the solvent and the solute, but it is employed beneficially and does not adversely affect the properties of the solute.

A further requirement is that the solute should be reasonably readily recovered from the extract (i.e., the solution after the extraction process) in those cases where the desired product is the solute. As figure indicates, merely reducing the pressure suffices to lower the solubility of caffeine in SC-$CO_2$. If the extract is saturated at high pressure, then the pressure can be reduced, caffeine will crystallize from the solution, and the depleted solvent can then be re-pressurized for reuse. In some cases the solute is not thermally labile, and the solvent can be recovered and separated from the solute by distillation. A further possibility is that the recovery of the solvent can take place through extractive distillation. For example, in the recovery of acetic acid from dilute aqueous solutions with methyl tert-butyl ether (MTBE), the extract contains both water and acetic acid; however, when the solvent is distilled, an azeotrope of water and MTBE is formed and the solute (acetic acid) is recovered as anhydrous glacial acetic acid. The azeotrope is then cooled, when it separates into an aqueous and an MTBE layer, and MTBE can be recycled directly.

A further consideration in the choice of solvent is that it should readily be separated from the raffinate (i.e., the feed material following extraction). Generally density differences suffice to bring about a high degree of separation, although in some cases the difference is so small that centrifuges must be employed. Centrifuges are also employed when the raffinate is a pulp that tends to entrain the extract. It may be necessary to wash such pulps with fresh solvent to recover trapped extract, particularly if a high yield of the solute is required.

The solvent will dissolve to some extent in the raffinate. One seeks to minimize the quantity lost in this way because it represents an economic loss, but possibly more importantly because it may contaminate the raffinate. The choice of a solvent with a very low solubility in the raffinate will naturally assist, but other considerations may force the choice of a solvent that has a significant solubility over one with a very low solubility. Removal of dissolved solvent from the raffinate may rely on the vapor pressure of the solvent being sufficiently lower than that of the raffinate that vapor scrubbing or distillation can remove it. Alternatively, if the raffinate is reasonably liquid, residual solvent may be stripped by adsorption on a solid such as activated carbon or clay. In extreme cases, it may be necessary to employ a second solvent that has a very low solubility in the raffinate to remove the solvent that was first used to remove the solute.

## Engineering Considerations

## Batch Operation

In the simplest extraction step, a feed is mixed with a solvent for a period sufficient for the solute to reach equilibrium between the two phases. The two phases are then separated, and the solute is recovered from the solvent. The question is how much of the solute will be extracted.

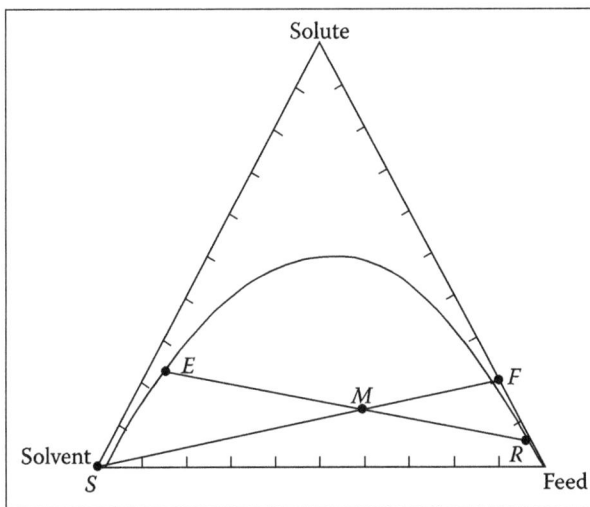

Extraction in a single stage.

A simple mass balance with feed F, solvent S, extract E, and raffinate R, and $c_i$ being the concentration of the solute in the ith stream, gives:

$$F + S = E + R = M$$
$$Fc_f + Sc_s = Ec_e^* + Rc_r^*$$

Where c* represents the concentration at equilibrium. This is illustrated in figure above.

The position of M on line SMF is found from the relationship,

$$\frac{F}{C} = \frac{\overline{MS}}{\overline{FM}}$$

where overstrike indicates the length of the section of the line. The equilibrium relationship, shown as the tie line EMR in figure above, links $c_e^*$ and $c_r^*$ such that at any one temperature T. For many systems, the distribution coefficient, $D_T$, is close to constant at low concentrations, but will vary with temperature.

$$c_e^* / c_r^* = D_T$$

If the solvent is recycled free of solute ($c_s = 0$), the quantity of solute extracted will be given by $Ec_e^*$ and the fraction extracted by $Ec_e^* / Fc_f$. Because the solvent is usually chosen to be reasonably insoluble in the feed, the volume of extract will be close to the volume of solvent fed (i.e., $E \approx S$), so that the fraction extracted will depend strongly on the volume ratio of feed to solvent, S/F. However, increasing the volume ratio will reduce the concentration of the solute in the extract, which will in turn increase the cost of removing the solute from the solvent. Therefore, there is an economic balance to be struck that sets a limit on the maximum quantity of solvent that can be employed and the maximum achievable recovery of the solute.

## Differential Batch Operation

It is possible to maximize the recovery of a solute by repeated extraction with fresh solvent. A practical way of achieving this is shown in Figure.

The feed and solvent are mixed together, then overflow to a settler and allowed to separate. The raffinate is returned to the mixer and the extract passes to a solvent recovery stage where the product is removed and the recovered solvent is recycled to the mixer. This type of process is employed where a highly valued solute is at low concentration in the feed. It is widely used in the essential oils industry for extracting trace aromas or flavorants from plant material. Particular care must be taken to avoid trace contaminants in the fresh solvent, as there is a risk that these will also be concentrated in the product and in turn contaminate it unacceptably.

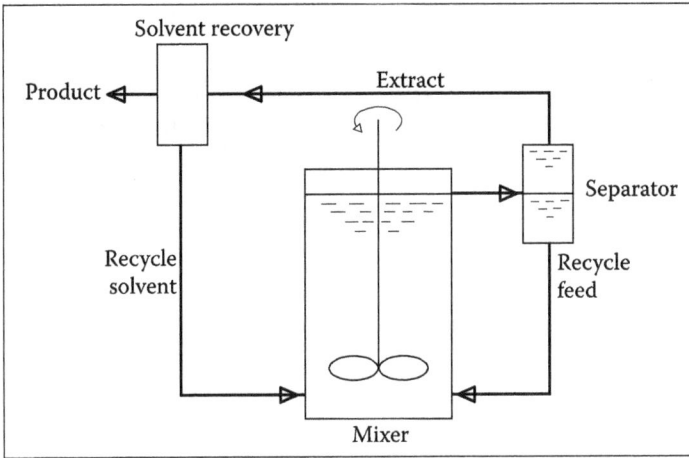

Differential extraction circuit.

## Countercurrent Operations

## Batch Countercurrent

In the process shown in figure, the quantity of solvent that must be separated from an increasingly dilute extract becomes very large. An alternative, which has the advantage of reducing the quantity of solvent for a given duty, is to operate in such a way that the fresh feed is extracted with solvent containing nearly the maximum quantity of solute. The raffinate from that stage is then extracted with solvent containing even less solute, and the extract from that stage then forms the feed to the first stage. This can be done batch-wise, as illustrated in Figure.

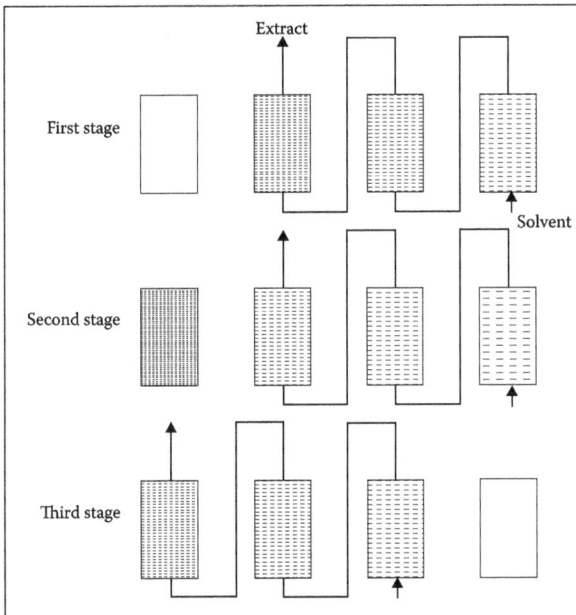

Batch countercurrent operation.

In figure, there are four tanks, the first of which is empty in the first stage of operations. Solvent enters the fourth tank, overflows to the third tank, which overflows in turn through the second. The extract comprises the solution leaving the second tank. The feed material is stripped of solute progressively from the second to the fourth tank.

In the second stage, the first tank is filled with fresh feed, while the material in the other tanks is progressively depleted. Eventually the material in the fourth tank loses all its solute, and the operation enters its third stage.

In the third stage of operations, the solvent enters the third tank and the extract leaves the first tank, while the raffinate is emptied from the fourth. The system is then effectively back at its starting condition, and the process continues.

The type of batch operation is frequently used where the feed material does not flow readily. It is somewhat expensive, partly because of the need to operate a large number of valves in a strict sequence, which involves control and maintenance challenges, and partly because the emptying of solids from the tank once the extraction is complete is not necessarily straightforward. Nevertheless, this type of extraction is quite widely applied in the industry.

## Mixer–settlers

Where the feed is reasonably fluid, then continuous countercurrent extraction is preferable to batch operation. In one variant, the feed and solvent are physically mixed, and then allowed to settle, usually under gravity. The settled extract phase then passes to the next mixer upstream while the settled raffinate passes to next mixer downstream.

The two flows are separated at the end of the settler by a simple weir arrangement.

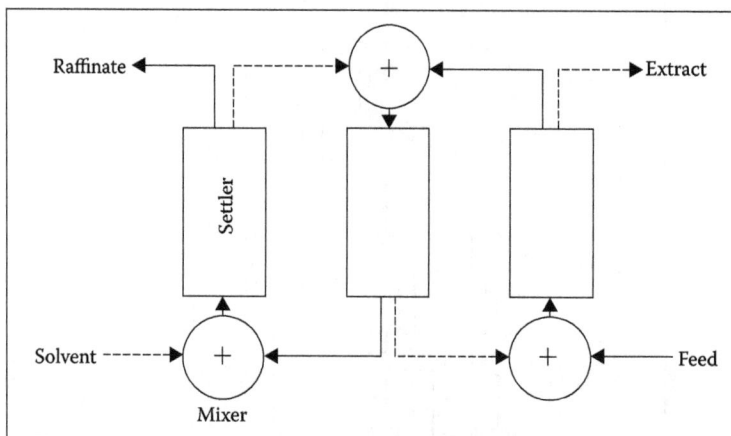

Continuous countercurrent mixer–settler.

The lighter extract phase overflows a weir at the end of the settler and is withdrawn. The heavier raffinate phase flows out at the bottom of the settler into a chamber, from the top of which it overflows. The height of the raffinate weir determines the depth of each phase in the settler, because a simple pressure balance in the settler and the raffi-nate chamber gives:

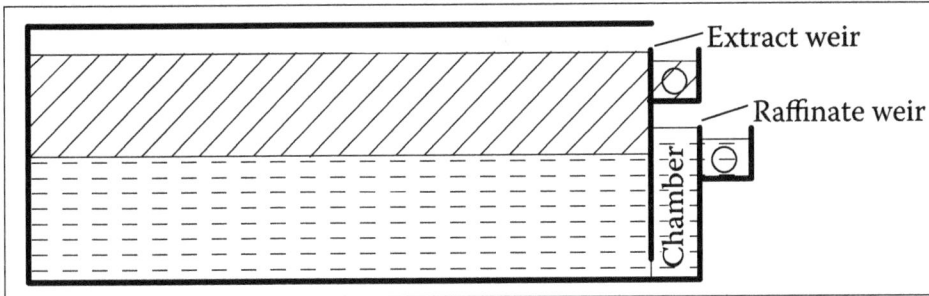

Elevation view of a settler, showing weir arrangement to separate phases.

$$h_{rs}\rho_r + h_{es}\rho_e = \text{pressure at base of settler} = h_{rc}\rho_r = \text{pressure at base of chamber.}$$

where h is the height, $\rho$ is the density of the phase, subscript r refers to raffinate, subscript e refers to extract, subscript s refers to settler, and subscript c refers to the chamber.

It is often found that any solids in the feed tend to collect at the interface between the extract and the raffinate in the settler, where they may interfere with the efficiency of separation. For this reason many settlers also have an arrangement by which material can be withdrawn from the region of the interface and treated separately for the recovery of the extract or raffinate.

The phases move from one mixer–settler unit to the next either by pumping or by gravity. Some designs incorporate a pump function in the mixer, which minimizes the number of pumps required. Each unit may be placed on a different level, so that one phase may gravitate while the other is pumped between stages.

Mixers require careful design. They have to disperse one phase in the other, and to do so without creating such fine droplets that the phases will not separate readily in the settler. Care must be taken to keep surface-active agents out of the system, as they will lower the interfacial tension between the phases and thus cause a fine dispersion that will not settle. Many biological systems contain natural surfactants; thus, it is often necessary in developing extraction systems to pilot them carefully to ensure that substances that affect the interfacial tension are not present—or, if they are present, to design the mixer to minimize the influence of the surfactant.

It is possible to employ centrifugal forces to hasten settling, and centrifuges are occasionally used instead of gravity settlers. There are designs of centrifugal mixer– settlers where several stages are packed within a single centrifugal unit. However, these tend to foul if there are any solids in the feed, and thus have not found widespread use in food processing.

## Columns

It is possible to contact two liquid phases countercurrently in a column.

In the simplest design, one phase may be dispersed in the other by spraying through nozzles. The droplets rise or fall (as the case may be) through a counterflowing stream of the other phase. In figure, the solvent phase is continuous, and droplets of the feed fall through a rising stream of solvent before forming a pool at the bottom of the column, from where the raffinate is removed. At the top of the column, the extract merely overflows.

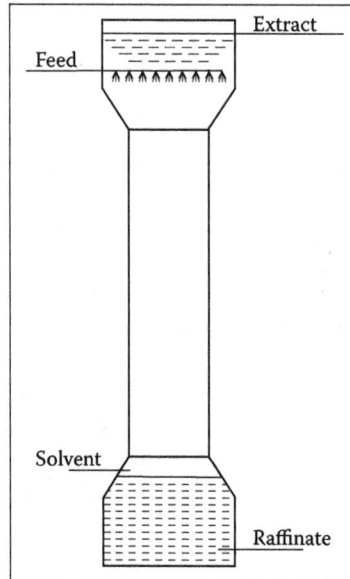

Extraction column operated with the solvent phase continuous.

This simple arrangement is not very efficient for a variety of reasons. One is that the surface area of the droplets is not very large. Another is that the falling droplets drag solvent molecules with them, and this means that solvent containing a lot of solute is mixed with lean solvent lower in the column. This reduces the concentration difference on which mass transfer depends, and thus reduces the efficiency.

For these reasons most columns contain a packing, the function of which is to increase the surface area of the phase that preferentially wets the packing. Because there is a larger area, the velocity of the dispersed phase is reduced and consequently there is less backmixing. There is a wide range of proprietary packing that may be employed. Some have less tendency than others to collect suspended solids, and are therefore preferred in food processing applications.

## Extent of Extraction

It is obviously necessary to be able to estimate how many stages of countercurrent extraction are necessary to achieve a desired degree of extraction. It may be necessary to remove essentially all of a particular solute from the feed. Alternatively, there may be little point in recovering the last traces of a solute if the value of those last traces is too low to justify the expense of building additional extraction stages and operating them.

Whichever the case, methods for estimating the number of stages required to perform a given extraction are essential.

With the equilibrium information available, it is obviously possible to solve the various mass balance equations numerically. However, for more than two or three stages, solving the resultant set of equations becomes tedious even in environments such as MATLAB, and, if anything, it is self-defeating, because it is not possible to achieve 100% efficiency. Thus, a simple estimate suffices in the majority of cases.

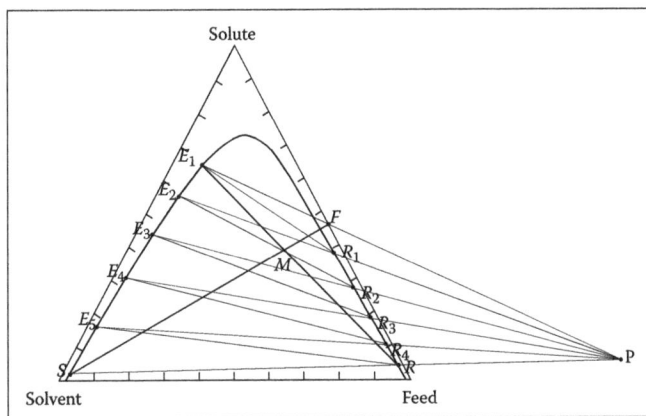

Graphical estimation of number of countercurrent stages.

For this reason graphical methods still find use. Figure gives an example. The feed concentration F and the solvent composition S should be known. Then assume a raffinate concentration R and an extract composition $E_1$. The extent of extraction obviously follows from $1 - (R/F)$. Construct SR and $E_1F$ and extend until they meet at P, the "operating point," because,

$$F + S = E_1 + R, \ F - E_1 = R - S = P$$

Then there is a tie line from $E_1$ to $R_1$, the raffinate concentration in equilibrium with $E_1$. A mass balance over the first and second stages gives,

$$F + E_2 = E_1 + R, \ F - E_1 = R - E_2 = P$$

Therefore, a construction from P through $R_1$ will give $E_2$, which in turn will give a tie line to $R_2$, and a construction from P through $R_2$ will give $E_3$. Continuing in this way, eventually the graphical $R_n$ will be at or below R. In the example given in Figure, $R_6 \approx$ R; thus, six stages are needed to reduce the concentration of the solute from F to R. The ratio of solvent to feed is given as was the case in Equation by the position of M.

There is a special case when there is no mutual solubility between the extract and the raffinate. In this case, the equilibrium is much simplified and can be shown on rectilinear coordinates with, typically, the raffinate concentration of the solute on the ordinate and the extract concentration of the solute on the abscissa.

As before, the feed concentration F and the solvent composition S should be known, and a raffinate concentration R and an extract composition $E_1$ is assumed. Then, the equilibrium between the two phases is given by the curve shown. Construct an operating line AB where point A has the coordinates [F, $E_1$] and point B the coordinates [R,S]. Mass balance considerations show that the slope of this line is the same as the phase ratio (i.e., the ratio of volumetric flow of extract to the volumetric flow of raffinate). This is readily shown from mass balance considerations and, of course, only holds true provided the raffinate and extract are mutually insoluble.

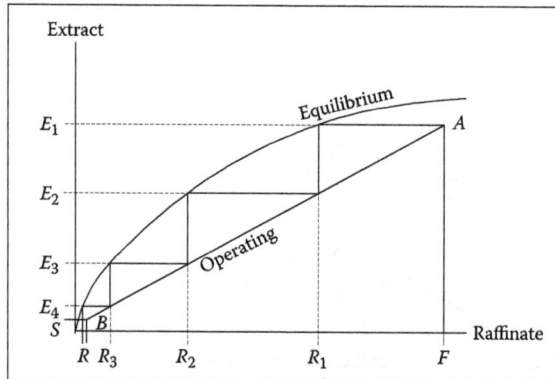

Graphical estimation of number of countercurrent stages where extract and raffinate are mutually insoluble.

Then if equilibrium is attained in each stage, the feed at concentration F will be in equilibrium with the final extract at a concentration $E_1$; thus, the raffinate concentration will fall from F to $R_1$. In the next stage $R_1$ will be extracted to equilibrium with $E_2$, and the raffinate concentration will fall further to $R_2$. Continuing in steps in this way, there will come a point where the raffinate concentration is below the desired final concentration R. In the example given in figure, four stages suffice to reduce the concentration to the desired level. The graphical representation of countercurrent extraction in this manner is known as the McCabe–Thiele diagram after the chemical engineers who originally derived it.

Note that if the flow of feed is reduced—that is, the phase ratio is increased—then the slope of the operating line will be increased. It is not possible to reduce the feed flow indefinitely—at some point the operating line will cross the equilibrium line, and point A will lie above the equilibrium. It is no longer possible to construct the diagram, and $E_1$ must be reduced until point A is again below the equilibrium line at the new phase ratio.

## Treatment of the Extract and Raffinate

### Extract

The extract will need to be treated to recover both the solvent and the solute. A range of separation technologies is available for this purpose. If the solute is not thermally labile, then separation may be effected by distillation, with recovery of the solvent in a condenser. Vacuum distillation can often be employed if the solute is thermally labile, although this may give rise to considerable cost because of the need to run the condenser at low temperatures to recover the solvent. It may also be possible to recover the solute by reducing the temperature of the extract and thus crystallizing the solute and settling, filtering, or centrifuging to remove the solvent from the solute crystals.

Chemical methods may also be employed. For instance, soy protein is extracted by alkaline water at approximately pH 9. Acidifying the water to about pH 4 precipitates the protein as a curd, which is centrifuged to remove the remaining water.

The solute can also be back-extracted from the extract. For instance, quinine is extracted from the bark of the cinchona tree into warm mineral oil and the solute is stripped from the extract using sulfuric acid–acidified water. The acidic water containing the quinine solute is filtered to remove insoluble material, and the quinine is recovered by adding alkali when the sulfate salt precipitates.

Similar principles can be applied to many extracts. The methods to be employed in any particular case will require testing before application, but in the majority of cases a simple and cost-effective method can be found for recovering solute and solvent separately. It is noteworthy that complete removal of the solute from the extract is not essential. The solvent may contain some of the solute and still be recycled to extract more material.

## Raffinate

The principal task in raffinate treatment is usually the complete removal of the last traces of solvent. In food processing, the raffinate often contains significant quantities of solids, which adds complications. However, considerable separation may be possible by agitating the pulp with water, when the solvent will coalesce above the aqueous layer and can be removed that way. Oily solvents can be removed by washing with a much lighter, volatile hydrocarbon such as hexane, from which the residual solvent can be recovered by distillation, while any hexane left in the raffinate is removed by heating.

A wider range of separation processes is available if the raffinate is liquid and contains minimal solids. In that case residual solvent may be removed by adsorption on activated carbon or even clay. Centrifuging will take advantage of the density difference to remove droplets that are too fine to settle under gravity. In some cases, fine air bubbles have been used to scavenge traces of solvent. The surface of the bubbles is hydrophobic, which therefore attracts the solvent, and can be removed from the surface as a scum.

## New Technologies

## General

Increasing energy costs and the global imperative to reduce the carbon footprint has sparked the development of a number of new separation techniques for the chemical, pharmaceutical, and food industries. Currently, many extraction processes in the food industry involve the use of organic solvents. However, these solvents not only present as atmospheric pollutants but also remain in the raffinate, as well as in the extracts, detracting from their purity. While water, not organic solvents, is most often used in the extraction of essential oils, water is also becoming a scarce commodity, sufficient to engender an interest in processes that conserve this precious solvent. Hence, to satisfy the growing demand for product purity, nonpolluting and energy-efficient processes, with the added advantage of using less solvent, alternative processes to the aforementioned solvent extraction methods are being sought. At present, supercritical fluid extraction (SFE) is the most extensively used alternative to solvent extraction with many commercially produced SFE compounds already available. More recent extraction techniques developed in the quest to create commercially viable, efficient, energy-saving, safe, compact, and sustainable extraction processes include extraction assisted by pulsed electric field, solvent-free microwave extraction (SFME), instant controlled pressure drop technology (DIC; from the French, Détente Instantanée Contrôlée), microwave hydrodiffusion and gravity (MHG), ultrasound assisted extraction, subcritical water extraction, high pressure–assisted extraction, aqueous two-phase extraction, and enzyme-assisted aqueous extraction.

## Supercritical Fluid Extraction

Owing to its low cost and ready availability at high purity, the mostly commonly used supercritical fluid solvent in food applications is carbon dioxide ($CO_2$). $CO_2$ is not flammable, with a moderate critical temperature (31°C) and pressure (7.4 MPa), ensuring its safety in handling, while its easy removal from the extract to sound physiological levels brokered its "generally regarded as safe" status—that is, it can be used in food processing without declaration. Moreover, when recycled during the process—for example, after recovery by reducing the pressure during caffeine extraction—it does not contribute to the carbon footprint. Some examples of everyday products where SC-$CO_2$ extraction is used are decaffeinated coffee and tea, flavor-enhanced orange juice, dealcoholized wine and beer, defatted meat and French fries, beer brewed with $CO_2$ hop extracts, rice parboiled using $CO_2$, spice extracts, and vitamin E– and β-carotene–enriched natural products. Until recently, SFE was considered too costly for producing low-value, high-volume commodity oils, and only viable if applied to high-value, low-volume specialty oils. However, increasingly stringent environmental regulations, particularly regarding the use of hexane, has led to significant progress in optimization of design and operation of large-scale supercritical oil extraction plants, such that their cost structure is now comparable with that of conventional plants. However, to realize the full potential of SFE in terms of oil extraction, it should be extended to include oil refining, as well as further extraction of valuable biomass components (e.g., proteins and carbohydrates). Hence, SC-$CO_2$ extraction, combined with subcritical and supercritical water extraction, may result in biorefineries that are "green" in the true sense of the word.

## Pulsed Electric Field-assisted Extraction

Pulsed electric field (PEF)-assisted extraction has been shown to enhance solid– liquid extraction processes in the food industry. PEF had been successfully employed to extract anthocyanins and phenolics in red wine must; sucrose, proteins, and inulin from chicory; betanine from beetroot, sugar beet, apple, and carrot juice; and maize germ and olive oils. The high extract yield, lower operating temperatures (preventing thermal degradation), high product quality and purity, high process efficiency, shorter extraction times, and lower energy cost of PEF extraction methods are key advantages of this technique, heralding its potential for energy-efficient and environmentally friendly food processing.

## Microwave-assisted Extraction

Essential oils, the single most widely extracted commodity, are extracted from herbs and spices and other botanicals for flavors, fragrances, and antimicrobial applications. Conventional extraction techniques, namely steam distillation, hydrodistillation, and extraction with lipophilic solvents, have been successfully replaced by SC-$CO_2$, with the attendant cost savings in terms of energy (lower process time and temperature), as well as superior product quality. The most recent developments in this field are the use of microwave-assisted hydrodistillation (MAHD), SFME, DIC, and MHG. Compared with hydrodistillation, MAHD, SFME, and MHG not only produce superior quality essential oils but also result in higher yields and significant savings in process time; in addition, energy savings and lower water consumption are achieved. The savings in energy and solvent is due to the absence of distillation or solvent extraction, those being the unit operations responsible for high energy and solvent consumption. MHG, in particular, shows great promise for industrial-scale operations. With both energy consumption and carbon emission being 6% that of hydrodistillation, as well as its very short process time, nonrequirement for water or other solvent nor for postprocess waste water treatment, and higher-purity final product, this process ticks all the boxes for a green, effective alternative to conventional solvent extraction techniques.

## Ultrasound-assisted Extraction

Ultrasound-assisted extraction (UAE) is another emerging technology with industrial-scale processing equipment designs available. Applications to food processes include extraction of vanillin, almond oils, herbal extracts,soy protein (with enhanced removal of flatulence-causing soluble sugars), polyphenols, and caffeine from green tea. UAE also offers a process that reduces the dependence on solvents such as hexane, with improved economics and environmental benefits (mainly due to an increased yield of extracted components), increased extraction rate, reduced extraction time, and higher process throughput. A more recent modification of UAE, namely ultrasound-assisted dynamic extraction, was used to extract chickpea oil. This process entails circulation of the solvent while the sample and solvent are subjected to ultrasound, and results in further reductions in solvent consumption and extraction time and, therefore, environmental impact.

## Subcritical Water Extraction

Subcritical water, also known as pressurized low-polarity water, or pressurized hot water extraction (PHWE), utilizes hot water (100–374°C) under pressure (1000–6000 kPa) to replace organic solvents. PHWE delivers higher extraction yields from solid samples than conventional solvents. This technique was used effectively to extract peppermint oil, carotenoids from microalgae, carnosic acid and aroma compounds from rosemary, quercetin from onion skins, and rice bran oil through simultaneous lipase inactivation, to name a few. Compared with conventional extraction methods (i.e., solid–liquid extraction, hydrodistillation, and organic solvents), PHWE offers several advantages, namely shorter extraction times, higher-quality extracts, a less costly extracting solvent, and an environmentally friendly process.

## High Pressure–assisted Extraction

High pressure–assisted extraction has also gained ground as an environmentally friendly alternative to solvent extraction. Advantages include shorter extraction times, higher yields, extract purity, and lower energy consumption. A variation on this, known as DIC, uses high-pressure steam to extract essential oils, followed by rapid transfer and cooling to a vacuum chamber. The rapid condensation in the vacuum tank produces a microemulsion of water and essential oils.

## Aqueous Two-phase Extraction

Aqueous two-phase extraction was successfully employed for purification and concentration of betalains (a natural colorant from beetroot), resulting in a simpler and therefore more environmentally friendly process.

## Enzyme-assisted Aqueous Extraction

This process has been applied to extract oils from various oil seeds and some fruits. The advantages of the method reside in the fact that processing occurs at relatively low temperatures and use water as a solvent, ensuring superior product quality and making the method safe and environmentally friendly. The presence of food-grade enzymes improves the oil yield.

## Impact of Refinin

Another aspect of vegetable oil processing is that the steps after solvent extraction, namely solvent recovery and refining, also have high-energy demands, consume large quantities of water and other chemical reagents, and produce significant quantities of effluent. The above concerns had been addressed by employing membrane technology, particularly ultrafiltration and nanofiltration. The technology has been applied successfully at the pilot scale for solvent recovery from soybean and cotton seed oil and shows great promise with regard to deacidification and degumming, obviating the use of sodium hydroxide. It had been estimated that using membrane technology for solvent recovery rather than heating could effect a savings of $2.1 \times 10^{12}$ kJ year$^{-1}$ in the United States alone, while having considerably less noxious effect on the environment.

## Centrifugation and Sedimentation

Sedimentation and centrifugation are used to separate immiscible liquids and solids from liquids. The separation is carried out by the application of either natural gravity or centrifugal forces.

## Field of Application

Centrifugation is typically used in the dairy industry in the clarification of milk, the skimming of milk and whey, the concentration of cream, in the production and the recovery of casein, in the cheese industry, and in lactose and whey protein processing, etc. This processing technique is also used in beverage technology, vegetable and fruit juices, coffee, tea, beer, wine, soy milk, oil and fat processing/recovery, cocoa butter, and in sugar manufacturing etc.

Centrifugation is used to separate mixtures of two or more phases, one of which is a continuous phase. The driving force behind the separation is the difference in density between the phases. By using centrifugal forces the separation process is accelerated. The necessary centrifugal forces are generated by rotating the materials. The force generated depends on the speed and radius of rotation. In raw milk for example, the skimmed milk is the continuous phase, the fat phase is a discontinuous phase formed of fat globules with diameters of some microns, and a third phase consists of solid particles, hairs, udder cells, straw etc. When the differences in density are large and time is not a limiting factor separation can take place by gravity (known as sedimentation and skimming). In beer production, clarification of the hot wort is carried out in order to remove particles (hot trub) to get a clear wort. The commonly used equipment for wort clarification is the whirlpool, where wort trub particles are separated in tangential flow.

## Separation by Gravity

- Batch-wise: this occurs in a vessel containing a dispersion of solid particles with a higher density than the liquid. In time these heavier particles fall to the bottom of the vessel. If the height of the vessel is shortened and the surface increased, the sedimentation time can be reduced.

- Continuous: the liquid containing the slurried particles is introduced at one end of the process and flows towards an overflow. The sedimentation capacity of the vessel can be increased by adding baffle plates (horizontal or inclined).

## Separation by Centrifugal Force

Centrifuges are classified into three groups:

- Tubular/disc bowl centrifuges for separating of immiscible liquids.
- Solid bowl/nozzle valve discharge centrifuges, for clarifying liquids by the removal of small amounts of solids.
- Conveyor bowl/reciprocating conveyor centrifuges, for dewatering sludges (with a high solids content).

## Disc Bowl Centrifuges

A tubular bowl centrifuge consists of a vertical cylinder, which rotates between 15000 – 50000 rpm, inside a stationary casing. It is used to separate immiscible liquids, e.g. vegetable oil and water or solids from liquid. The two components are separated into annular layers, with the denser liquid solid setting nearer to the bowl wall. The two layers which are then discharged separately. Typically, the disc bowl centrifuge is more widely used in the food industry as it can achieve a better separation due to the thinner layers of liquid formed. With the disc bowl centrifuge, the cylindrical bowl contains inverted cones or discs. The liquids only have to travel a short distance to achieve separation. These centrifuges operate at 2000 – 7000 rpm and have capacities of up to 150000 l/h. Disc bowl centrifuges are used to separate cream from milk, and to clarify oils, coffee extracts and juices or to separate starch from slurry.

## Solid Bowl Nozzle

A solid bowl centrifuge is the simplest solids/liquid centrifuge and is useful when small amounts of solids must be removed from large volumes of liquid. It consists of a rotating cylindrical bowl. Liquor is fed into the bowl; the solids settle out against the bowl wall whilst the liquid spills over the top of the bowl. Periodically the centrifuge has to be stopped to enable the cake to be removed. Liquors containing higher levels of solids, i.e. >3 % w/w, can be separated using nozzle or valve discharge centrifuges. These centrifuges are a modified disc bowl centrifuge with a double conical bowl and enable the discharge of solids automatically. These types of centrifuges are used to treat oils, juices, beer and starches to recover yeast cells. They have capacities of up to 300000 l/h. A special type is the "bactofuge", which is specially designed to separate micro-organisms from milk. Bacteria, and particularly spores, have a higher density than milk and the solids are called bactofugate.

## Reciprocating Conveyor

These centrifuges are used when the feed contains high levels of solids (sludges). They are used, for example, to recover animal and vegetable proteins (i.e. precipitated casein from skimmed milk), to separate coffee, cocoa and tea slurries and to desludge oils. In the conveyor bowl centrifuge (decanter), the solid bowl rotates at 25 rpm faster than the screw conveyor. This causes the solids to be conveyed to one end of the centrifuge whereas the liquid fraction moves to the other larger-diameter end. The reciprocating conveyor centrifuge is used to separate fragile solids (e.g. crystals from liquor). The feed enters a rotating basket through a funnel, which rotates at the same speed. This gradually accelerates the liquid to the bowl speed and thus minimises shear forces. Liquid passes through perforations in the bowl wall. When the layer of cake has built up it is pushed forward by a reciprocating arm. The basket centrifuge has a perforated basket lined with a filtering medium, which rotates at 2000 rpm. Separation occurs in cycles, which last from 5 – 30 minutes. In the three stages of the cycle the feed liquor first enters the slowly rotating bowl, the speed is then increased and separation takes place, finally the speed of the bowl is reduced and the cake is discharged through the base. Capacities for this group of centrifuges are up to 90000 l/h.

## Filteration

Filtration is the separation of solids from a suspension in a liquid by means of a porous medium, screen or filter cloth, which retains the solids and allows the liquid to pass through.

## Field of Application

Filtration is used in the food, drink and milk industry to fulfil the following functions:

- To clarify liquid products, by the removal of small amounts of solid particles (e.g. for wine, beer oils and syrups). The objective is to recover the filtrate in this operation.

- To separate a liquid from a significant quantity of solid material, where the overall objective of the operation is to obtain the filtrate or cake, or both, is (e.g. for fruit juices or beer).

Filtration equipment operates either by the application of pressure (pressure filtration) to the feed side or by the application of a vacuum (vacuum filtration) to the filtrate side. The two main types of pressure filtration are the plate and frame filter press and the leaf filter.

## Plate and Frame Filter Press

This type of filter consists of plates and frames arranged alternatively and supported on a pair of rails. The hollow frame is separated from the plate by the filter cloth. The slurry is pumped through a port in each frame and the filtrate passes through the cloth and flows down the grooved surfaces of the plates and is drained through an outlet channel in the base of each plate. The filter operates at a pressure of between 250 to 800 kPa. The filter press is operated batch wise; the optimum cycle time depends on the resistance offered by the filter cake and the time taken to dismantle and refit the press. A special type of "plate and frame" filter press is the "membrane" filter press. A membrane is mounted on the plate which can be pressurised with air or water. Due to the higher pressure (up to 20 bar) on the filter cake, more liquid is recovered resulting in a dryer filter cake. Filter presses are available where the dismantling, emptying and refitting of the press are carried out in a semi–automatic manner. Sometimes, filter aids such as (perlite or diatomaceous earth) are used as a pre-coat or body feed to improve the filtration. The equipment is reliable and easily maintained and is widely used, particularly for the production of apple juice and cider and in edible oil refining (bleaching).

## Leaf Filter

The need to develop much larger capacity units led to the introduction of the leaf filters. These consist of mesh "leaves" which are coated in filter medium and supported on a hollow frame, which forms the outlet channel for the filtrate. The leaves can be stacked horizontally or vertically. Feed liquor is pumped into the shell at a pressure of approximately 400 kPa. When the filtration is completed, the cake is blown or washed from the leaves.

## Vacuum Filtration

Vacuum filters are normally operated continuously. Liquor is sucked through the filter plate/cloth and a cake of solids is deposited on the cloth. The pressure difference on the downstream side of the filter plate is normally limited to 100 kPa due the cost of vacuum generation. Two common types of vacuum filter are the rotary drum filter and the rotary disc filter. Sometimes, filter aids are used as a pre-coat or body feed to improve filtration. In these cases, a knife is used to scrape off the cake. Rotary drum filters consist of a slowly rotating cylinder, which is divided into compartments, which are themselves covered with a filter cloth and connected to a central vacuum pump. As the drum rotates it dips into a bath of liquor. The filtrate flows though the filter cloth of the immersed compartment. When the compartment leaves the bath the filter cake is sucked free of liquor and washed. As the drum rotates further the vacuum is released for the compartment in question and the cake is loosened from the cloth by the application of compressed air, and then removed by means of a scraper. The same procedure occurs for each compartment in turn as the cycle is repeated. Rotary vacuum discs filters consist of a series of vertical discs which rotate slowly in a bath of liquor in a similar cycle to drum filters. Each disc is divided into segments and each segment has an outlet to a central shaft. The discs are fitted with scrapers to continuously remove the cake.

## Membrane Seperation

Membrane technology has been used in specialized applications in the food industry for more than 30 years. The technology can be applied to several production methods, including milk-solids separations in the dairy industry, juice clarification and concentration, concentration of whey protein, sugar and water purification, and waste management. Several filtration mediums exist as well as many types of membrane configurations. Knowledge of the various membrane technologies and how they are used in the food industry can enhance overall production and offers cost-cutting options for a variety of separations.

## Membrane Basics

Separation is based on principles that rely on the chemical and physical properties of particles and molecules. For example, centrifugation uses the physical property of weight to separate solids from liquids. Another example, ion exchange, relies on the principle of charge to separate different species from one another. Other principles such as vapor pressure, solubility and diffusion also can perform separations. Membranes use the principle of size to separate different materials.

Membrane filters are very thin microporous sheets of film attached to a thicker porous support structure. At its most basic, a membrane serves as a sieve, separating solids from liquids forced through it. Not only can membranes separate solids from liquids, they can separate soluble molecules and ionic particles of different sizes from each other.

Membranes in the process industry use tangential flow filtration, or crossflow filtration. In tangential flow filtration (TFF) the flow of the feed stream runs parallel to the surface of the membrane at high velocities.

As the fluid passes across the membrane surface it acts to sweep the retained components, or retentate, from the membrane surface and back into the bulk solution to avoid plugging the pores. This is opposed to traditional through-flow or dead-end filtration, in which the feed stream flows perpendicular to the surface of the membrane, the object of which is to form a dry cake. In TFF, the fraction that contains the solutes and lower molecular weight components that can pass through the membrane is known as permeate. The retentate is continuously recirculated, passing over the membrane surface until the desired effect—such as the concentration or clarification of a desired product—is achieved. The rate of permeation is known as flux and gives an indication of membrane performance.

TFF operations can perform both concentration and clarification applications. In concentration, the membrane retains the desired product, and liquid is removed as permeate. The retentate becomes more and more concentrated as permeate is removed. In clarification applications, the desired product passes through the membrane and is collected as permeate, perhaps leaving insoluble materials or other undesired compounds in the retentate. Both concentration and clarification operations are used extensively in the food industry, primarily to process juice and other beverages.

## Membrane Construction

Membranes are fabricated from many different types of materials. Initially, reverse osmosis and ultrafiltration membranes were cellulose-based, but are now made from polymers based on engineering plastics. Typical materials have included polyacrylonitrile (PAN) and blends of PAN with polyvinyl chloride (PVC), aromatic polyamid, and polyvinylidene fluoride (PVDF). Today, most polymeric membranes used in the food industry are made of polysulfone (PS) or polyethersulfone (PES). Materials also are chosen for their cleanability and ability to withstand a variety of conditions under which it might perform.

Membranes can be divided into four basic categories: reverse osmosis (RO), nanofiltration (NO), ultrafiltration (UF), and microfiltration (MF). Each of these categories is distinguished by the size of the species they retain. Retention is based on the pore size of the membrane.

Reverse osmosis has the tightest pore construction, and can separate in the ionic range. Nanofiltration, a newer category of membranes, operates similar to reverse osmosis but has a somewhat looser construction, allowing monovalent ions and some divalent ions to pass. Ultrafiltration is used to separate different-size molecules such as proteins and other macromolecules. Microfiltration membranes have the largest pore sizes of all categories, and is primarily used for removing suspended solids and bacteria. Another difference between these types of filtration is the pressures at which they operate. RO membranes operate up to 1,500 psi, NO membranes operate up to 300 psi, UF membranes operate from 10 to 200 psi, while MF membranes operate in the range of 1 to 25 psi. Pore size differences between the membranes determine the operating pressure. Higher pressures are necessary to force liquid through membranes with smaller pore sizes.

## Filtration Foundation

The module design—or support structure—of a membrane is critical to its performance. Some factors to consider include flux (the rate of permeation), the solids content of the process fluid, cost, cleanability and scalability. Food industry applications make use of four basic module designs: spiral-wound, tubular, hollow-fiber, and plate-and-frame styled systems.

Spiral-wound membranes cover more than 60% of food-industry applications, mainly for dairy and other soluble protein processing, polysaccharide gum concentration, and in most RO and NO applications. Figure shows a schematic of the spiral-wound design. Fluid is pumped into the spacer channel parallel to the membrane surface and permeate passes through the membrane and into a porous permeate channel until it reaches a perforated tube at the center which acts as a permeate carrier.

Although resistance to fouling is good, these membranes have difficulty handling viscous material, or material with a high solids content.

Tubular systems account for about 10% to 15% of food industry applications. Tubular designs have a porous outer structure with a semi-permeable membrane coating on the inside of the tube. the module consists of a collection of tubes fastened together at each end and encased in a module.

The shell of the module collects permeate while retentate discharges at the end. Tubular designs are easy to clean and can be visually inspected. They can handle liquids with high solids content and larger suspended particulates better than some of the other membrane designs. The membrane area, however, is typically small. Tubular membranes are suitable for beverage clarification or the reverse osmosis of pulp containing juices.

Plate-and-frame and hollow-fiber systems are among the miscellaneous configurations that make up the remaining percentage of designs used in food industry applications. In plate-and-frame styles, flat-sheet membranes are affixed to both sides of a porous plate and sandwiched in a holder. the feed stream enters the system and is directed via several channels to sweep over the surface of the membrane.

The fluid flows from these channels into the same outlet line and leaves as retentate. The permeate passes through the membrane into channels separate from the retentate and leaves as permeate. Several membranes and their supports can be stacked together in the holder to increase the overall membrane surface area. The advantage of this system is that if one membrane fails, it typically can be replaced at low cost. Also, plate-and-frame systems offer increased diversity. Once the initial capital cost to acquire the hardware is accomplished, a variety of membranes may be used in them. For example, both ultrafiltration and microfiltration processes may be done on the same unit merely by swapping out the proper membranes. Plate-and-frame systems have been used for the dealcoholization of beer in Europe and Australia, and also have been used for some high-viscosity concentration applications in the dairy industry.

Hollow-fiber membranes are similar to tubular membranes except that the hollow fibers are much smaller. The inside diameter of the fiber may range from 0.5 to 1.1 mm as opposed to 12.5 to 25 mm for the tubular design. The feed stream flows through the inside of the fibers and the permeate is collected in the containment shell. Hollow-fiber elements may have of hundreds of fibers all oriented in parallel. This allows for a high packing density and resistance to blockage of the flow channels. Also, this membrane can be backwashed to aid in cleaning. The strength of the fibers, however, is a limitation and low transmembrane pressures must be used to avoid bursting the fibers. The entire element must be discarded if even one fiber breaks.

## Reverse Osmosis

Reverse osmosis has become a standard process in the food industry. It is used to purify water for plant operations, to concentrate cheese whey proteins or milk in the dairy industry, for sugar concentration in the cereal processing industry, for concentration of juices, and for wastewater treatment in meat and fish processing industries. Reverse osmosis deserves a special look because of its suitability for a wide variety of applications.

Natural osmosis is a phenomenon in which a liquid passes through a semi-permeable membrane from a dilute solution to a more concentrated one. In reverse osmosis, pressure is applied to the more concentrated solution forcing water to flow towards the dilute solution.

In this process, the reject water is typically 30% to 50% of the feed flow. Thus, at maximum efficiency, every 100 gallons of water entering the system would produce 50 gallons of purified water. RO is helping to save companies both energy and water by helping to limit evaporation steps or by treatment of wastewater streams.

Many foods require the removal of large amounts of water to concentrate the product for more efficient packaging or shipping. Although evaporation is common, it requires substantial amounts of energy requirements compared with RO. In the U.S. the corn wet-milling industry alone consumes about 93.7 trillion Btu/year, which is equal to 90% of the energy consumed in grain milling operations. The amount of water evaporated in this industry is around 35 billion lb/year for steep water, and around 12 billion lb/year for sweet water. The energy required for RO is approximately 110 kJ/kg water versus 700 kJ/kg for the most efficient evaporator, resulting in substantial savings.

Wastewater is a problem in any industry. Reducing the level of the dissolved solids and biological oxygen demand (BOD) is sometimes the only way processing plants can discharge water safely. Also, removing dissolved solids from water allows it to be reused, which not only cuts down on water usage but on the amount that is discharged. The collected solids also may be recovered if they are of value. For example, an RO system can recover proteins, sugars, and enzymes from wastewater that may be reused within plant operations.

Membrane technology has made a tremendous impact on the food industry over the last several years. The separation of materials for different applications has become an important industrial operation. Considerable progress continues to be made in mem brane technology, and newer applications for existing systems are being discovered as the trend is to create integrated systems which utilize several different membrane types within a process.

# Size Reduction

Size reduction is a unit operation involving such activities as cutting, slicing, milling or pulping of food. It requires energy input to overcome a breaking stress. When this happens the food will break along lines of weakness resulting in release of energy and sound. So think of the last time you ate an apple, that sound that you heard is energy being given off. Where enough energy is not added to the material for breaking, it returns to its original size. This is a characteristic of elasticity. Therefore to break materials, the energy input must overcome both the elastic stress limit and breaking stress.

The three types of forces involved in size reduction of solids are compression, impact and shear. Compression is a grinding force, like when you pulverize cereal in a mortar. Impact occurs when material is thrown against a surface causing it to disintegrate. For example, if you take an egg and throw it against a wall causing it to splatter, that's an example of impact. Shear is a slicing force where you have one surface sliding over another. The typical example of this is the use of knives to slice food. For example, slicing bread. All of these forces normally exists together although one usually predominates. For example, when you slice bread, the primary force is shear, but there is also compression as the knife is forced downward on the bread and impact, as the knife is first brought down to make contact with the surface of the bread.

Equipment used in size reduction of solids are generally placed into three categories:

1. Cutting, slicing, dicing, mincing, shredding and flaking equipment.

2. Pulping equipment.

3. Milling.

Size reduction of meat, fresh fruits and vegetables is generally done using equipment from the first two categories while grains, pulses, spices and other low-moisture foods are processed by milling.

## Energy Requirement

The amount of energy applied in milling will depend on the friability of the food (ability to fracture), moisture content and heat sensitivity. A food material's friability will be affected by its internal structure. For example, fibrous roots such as ginger will be harder to pulp compared to carrots which is less fibrous and much softer. Dehydrated cooked beans containing gelatinized starch will be much easier to mill compared to uncooked beans that is crystalline and tough. In general, harder foods will have fewer lines of weakness and will therefore need longer residence time, larger machines, or more abrasiveness. This will in turn use up more energy.

Reduction of energy consumption may require adjustment in moisture for some dried foods. For example, wheat is conditioned by adjusting the moisture content to a higher moisture. This results in a softening of the endosperm, making it easier to be extracted with little bran contamination. Of course there has to be a delicate balance since too much moisture could make a sticky mess rather than efficient grinding and separation. On the other hand, grinding at loo low a moisture content can create excessive dust and fire risks.

Heat-sensitive foods such as spices which are highly aromatic and can lose flavor during processing, should be heated under cool conditions. This may involve pumping a coolant through the mill to remove heat as it is generated.

Determination of energy required in size reduction is generally done using either Kick's law or Rittinger's law. Kicks's law is used for coarse grinding when the size ratio of original material to end materials is 8:1. Bond's law and Retinger's law are used for intermediate and fine grinding respectively, where ratios approach or exceed 100:1. Kick's Law says:

$$E = K_k \ln\left(d_1 / d_2\right)$$

Where, E ($Wh^{-1}kg^{-1}$) is energy required; $K_k$ is Kick's constant; $d_1$ is initial particle size, and $d_2$ is end product particle size.

## Effect of Size Reduction

Advantages: Size reduction provide great benefits. In the process, foods are reduced to a size that is convenient, preferred or necessary for processing. For example, materials can be more easily transported at a lower cost due to reduction in bulk density. Efficiency in mixing is improved along with heat transfer, and extraction of food components such as flavors and oils. Consumers experience improved sensory attributes, especially texture and appearance. For example, a customer may prefer finely milled corn to make their porridge compared to corn grits; or turkey slices in their sandwich instead of minced turkey.

Disadvantages: Size reduction, for example milling, may cause loss of volotiles, especially when working with heat sensitive foods like spices. In wheat milling, important food components may be lost due to separation of the bran and aleurone layer which contains proteins, vitamins and minerals. For this reason wheat flour is fortified to replace loss of micronutrients. The increased surface area resulting from size reduction could lead to increased oxidation and loss of some nutrients such as vitamin A and essential fatty acids. Enzymes may also be activated, leading to faster spoilage and reduction in shelf life. The use of dull knives could lead to damage of cell structure and increased juice-drainage and nutrient loss in fruits and vegetables. Size reduction does not improve food safety but may increase risk of food contamination due to opening surface area. Skin, husk and bran surfaces of food that are highly contaminated could cause contamination of the internal food matrix. Like any other process, size reduction on unsanitized surfaces will increase food safety risks.

# Coating of Food

Fresh and minimally processed foodstuffs like fresh-cut produce, meat, fish as well as ready-to-eat meals with fresh components (e.g. vegetables, herbs) are important market sectors in retail food industry. However, because of their fresh nature, mild processing technologies or increased surfaces as consequence of cutting processes, these products are very sensitive to food quality changes. Deterioration can occur by chemical, physical as well as microbiological processes like water loss, enzymatic based and light induced colour changes (e.g. browning), oxidation, loss of cellular integrity (softening) or a growth of microorganisms. Relevant for food safety are foodborne pathogens, especially psychotropic bacteria like Listeria monocytogenes, which can grow in the cooling chain and species of the family Enterobacteriaceae like Salmonella enterica or Escherichia coli. Current prevention strategies are versatile and include distribution in the cooling chain, modified atmosphere packaging, controlled atmosphere storage or the application of preservatives. Functional edible coatings could be an interesting approach for overcoming some of the mentioned quality losses.

## Functional Edible Coatings

Edible coatings are consumable films, which enable supporting structures and protective layers on food surfaces. By selection of suitable matrices and incorporation of functional additives food quality changes by moisture transfer, oxidation processes, respiration, loss of volatile flavors or microbial growth can be reduced or even prevented.

Various raw materials are suitable for the development of edible coatings: hydrocolloids based on proteins of animal or plant sources (e.g. whey, soy, corn, legumes) or polysaccharides (e.g. cellulose derivates, alginates or starches), lipids (e.g. waxes, shellac, fatty acids) or even synthetic polymers (e.g. polyvinyl acetate). Their selection depends on the requirements on barrier properties (water vapour, oxygen, carbon dioxide), mechanical strength, gloss and durability. Examples for functional components could be antimicrobials (e.g. organic acids, fatty acid esters, polypeptides, essential oils), antioxidants (e.g. ascorbic acid, oxalic acid, cysteine), texture improvers (e.g. calcium salts), aroma or nutraceutical compounds (e.g. vitamins).

The application of the coating can be realised by dipping or spraying methods followed by film formation processes like removing the solvent (normally water), coagulation of proteins, solidification of gel structures or chemical reactions. Relevant factors for the film formation are the chemical composition of the coating material (hydrophilic or hydrophobic), the concentration and viscosity, which influence the homogenous distribution, flexibility and the barrier properties (gas exchange) of the layer as well as the release of the functional component by diffusion.

Generally, the application of edible coatings on foodstuff has been known for centuries, for example the coating of fruits with waxes in early china in order to prevent water loss and structural changes.

There are also edible films for specific food products commercially available like sucrose fatty acids for fresh produce or collagen casings for sausages.

In the last years a lot of research has been conducted on advanced functional edible coatings for broader food applications with promising results.

For example, Rojas have reported that edible films based on alginate/ gellan and N-acetylcysteine as antioxidants can prevent the fresh-cut apple wedges from browning within a storage period of 21 days. The effects of essential oils as antimicrobial agents in alginate-based edible coatings on apple surfaces were investigated by Raybaudi-Massilia et al. They have observed that low concentrations of lemongrass and cinnamon oil can inactivate more than 4 log of Escherichia coli O157H7 on the surface and extend significantly the shelf life of cutted apples.

An essential aspect in this research field is the knowledge about the specific foodstuff properties (e.g. oxygen sensitivity, microflora, respiration) as well as their main deterioration pathways, which are a prerequisite for the development of effective functional coatings without adverse effects on the quality. Furthermore, the effects of the edible films on the internal gas atmosphere and their consequences on the food quality and sensory properties have to be determined. This underlines that such functional edible coatings have to be tailor made according to the requirements of each product.

## Mixing of Food

Mixing or dispersing ingredients is an essential step in most food processing operations. Ingredients can be used in any physical state of matter - solid, liquid or even gas. They can be dissolved to form a single uniform phase, or dispersed homogeneously to form multiphase mixtures. Today, a large majority of processed ready-to-eat products are indeed multiphase dispersions: solids dispersed in liquids (e.g., baked beans), emulsions (e.g., soups), or bubbly dispersions (e.g., ice cream, meringue, sponge cakes, etc.). Mixing is therefore recognized as one of the most widely practiced food processing operations. A wide variety of mixing equipment is in use, and novel mixing technologies are constantly emerging on the industrial scene. A glance at some of the recent issues discussed in food abstracts provides ample evidence of innovation occurring in this area, especially in the issues relating to mechanical features. Yet, it would not be inaccurate to describe food mixing as one of the least understood processing operations. Despite the wide existence of a variety of mixers effectively used thus far, the mixing mechanisms operating within these devices are hardly understood. Consequently, it is extremely difficult to relate operating conditions with product quality on a firm scientific basis; and process design and control can only be achieved with the help of expensive practical trials.

## Special Features of Food Mixing

- Mixing involves a whole spectrum of materials from dry free-flowing powders to thin, viscous liquids and viscous pastes such as dough.

- More often than not, mixing involves too many components existing in different physical states, which have widely differing and time-dependent properties.

- Energy requirements for dispersing each component can also differ widely. For instance, emulsification requires high energy, whereas dispersion of delicate particulate matter in shear sensitive liquids requires relatively lower energy levels (e.g., dispersion of nuts into chocolate or whole fruits into yogurt).

- Mixing particulates also involves other characteristics: mixing and segregation occur simultaneously, and particulates are often polydisperse. Particle-liquid dispersion either involves particles dispersed into liquid bulk, or relatively low amounts of liquids dispersed with high volumes of particles (e.g., dispersion of flavors). Furthermore, when particles are dispersed into liquids, the rheological and interfacial properties of the continuous phase can change as mixing progresses. Moreover, segregation of blended components can also occur during discharge from the mixer: design of discharge is therefore critical.

- Food mixing can involve dispersion of air or gas bubbles into liquids and pastes. Bubble incorporation in processes (e.g., the manufacture of ice creams, sponge cakes, meringues, and bubbly chocolate confectionery) is so widely practiced that air and gases are increasingly being recognized as "food ingredients".

- In contrast, bubble incorporation, which inevitably accompanies the mixing of viscous recipes (e.g., sauces and salad cream), is undesirable, since it can result in inconsistent filling of packages and acceleration in spoilage. De-aeration or bubble exclusion can be classified as a food mixing operation since the end product results in a greater level of homogeneity.

- An idiosyncratic feature of food mixing (like many other food processing operations) is that it is rarely a process where mixing is the sole intended effect. A multitude of physico-chemical processes occurs simultaneously in the mixer environment, and the effectiveness of mixing can only be assessed in the context of end product quality.

• Finally, the effect of mixing can continue well after the mixing action has ceased, and it could be quite some time before the end point is reached. In such situations, on-line monitoring and process control can be very challenging.

It is therefore evident that the state of any mixture is the result of several highly complex mixing mechanisms operating in parallel. At this stage, it is desirable to discuss the ways of describing mixtures.

## Assessment of Mixedness

## Scale of Scrutiny

Although mixing normally aims to achieve an almost uniform distribution of components, the degree of uniformity must be assessable. An obvious method is to measure the concentration of each component and express it in terms of appropriately defined relative concentration. A problem at once arises. Since such assessments depend on sample size, what then should the scale of scrutiny be? It is inevitable that, as the scale of scrutiny decreases, a given component will appear more segregated. This is illustrated in Figure. It is necessary that any practical process should achieve homogeneity on a pre-determined scale of scrutiny. From a producer's perspective, this scale may correspond to the volume of unit packages.

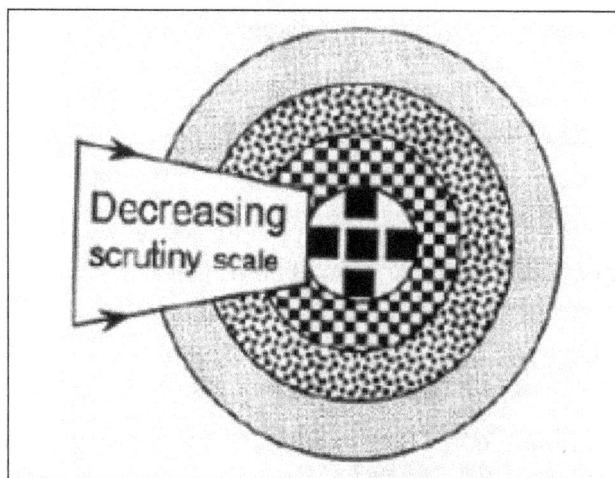

Appearance of mixture as scale of scrutiny decreases progressively.
It is evident that the components appear more segregated.

However, this does not necessarily correspond to the consumer's scale of scrutiny, which could be much smaller. One example used concerns the mixing of nutrients to form a cake for animal feeding. If the primary concern were to ensure that each animal receives the correct amount of nutrients daily, then the appropriate scale of scrutiny would be the volume corresponding to the daily intake of cake. However, if the criterion is to control the nutrient intake on a weekly basis, then the scale of scrutiny should be chosen according to the weekly consumption of cake. In the latter case, significant variations in nutrient concentration between daily feeds might exist, but averaged over the entire week, the nutrient intake would satisfy the overall requirement. It is therefore important to produce a product that is homogeneous based on the consumer's scale of scrutiny.

## References

- Cleaning-and-disinfection-improving-food-safety-and-operational-efficiency-in-food-processing, signature-series: foodsafetymagazine.com, Retrieved 18 April, 2019

- Sorting-screening-grading-dehulling-trimming-destemming, processing-technology: hyfoma.com, Retrieved 23 August, 2019

- Filtration- 97: safefoodfactory.com, Retrieved 29 July, 2019

- Membrane-separation-technology-in-the-food-in-0001: foodonline.com, Retrieved 25 March, 2019

- Principles-of-size-reduction-solids: cwsimons.com, Retrieved 19 May, 2019

- Functional-edible-coatings-for-fresh-food-products-2157: longdom.org, Retrieved 26 January, 2019

# Processing by Application of Heat

Heat processing is defined as the food processing technique that makes use of high temperatures to sterilize food. It includes blanching, pasteurization, evaporation, distillation, dehydration, smoking, frying, extrusion cooking, etc. All these methods related to heat processing have been carefully analyzed in this chapter.

## Heat Processing

Thermal processing is a commercial technique used to sterilize food through the use of high temperatures. The primary purpose of thermal processing is to destroy potential toxins in food. The process does have limitations and its application must be carefully overseen by an authority who understands the importance of variables in regulating thermal processing.

Thermal processing is a food sterilization technique in which the food is heated at a temperature high enough to destroy microbes and enzymes. The specific amount of time required depends upon the specific food and the growth habits of the enzymes or microbes. Both the texture and the nutritional content of the food may be altered due to thermal processing.

### Methods

Food may be sterilized using in-package sterilization techniques. Using this technique, the food is sterilized while it is already in a bottle, can or other package. The other option is UHT (ultra-high temperature) or aseptically processed products, which require the package and the food to be sterilized using thermal processing separately before they are sealed together.

### Acids

The presence of acids alters the temperature and processing times required for thermal processing. Some types of spores cannot reproduce in acidic environments, while acid helps aid in the destruction of other microorganisms.

### Processing Authority

The Food Safety and Inspection Service Canning Regulations require the presence of a processing authority to oversee the thermal processing. The processing authority is required to be an expert on food microbiology, thermobacteria, processing systems and heating characteristics of food.

## Limitations

In order to be classified as having commercial sterility, all the microorganisms do not have to be destroyed. Commercial sterility implies only that any remaining microbes will be incapable of continuing to grow and thrive in the food.

## Types of Heat Exchanger

Food and beverages are heat treated for several reasons, among which the most frequent and important is to inactivate microbial population and therefore stabilize and prolong shelf life. Heat transfer has to be rapid and effective, in order to avoid as much as possible any damage to nutritive and organoleptic qualities of food, and also to save time and cut down fuelling costs, all important achievement for the food industry. Computer programs are available to design heat exchangers to the best possible efficiency. In a heat exchanger, thermal energy is transferred from one solid or fluid to another solid or fluid. In designing the equipment, heat transfer equations are applied to calculate this transfer of energy in order to carry it out efficiently and under controlled conditions. Besides good heat exchanger design, other factors are of paramount importance to obtain heat transfer efficiency; among them is the avoidance of any debris (either mineral or organic) accumulation on the heat exchanging surfaces. For this reasons, if water is used as the heat exchanging fluid, it has to be deionized. Food fluids (e.g. milk, sauces, etc.) treated with the equipment have to be thoroughly removed at the end of every processed batch with adequate cleaning procedures. If exchanging surfaces are not perfectly clean, heat exchange efficiency will be compromised.

## Tubular Heat Exchangers

If one or both of the materials that are exchanging heat are fluids, flowing continuously through the equipment and acquiring/giving up heat, the process is very efficient, and these equipments are called "continuous flow heat exchangers". These equipment are often employed to pasteurize milk or other beverages, and the heat exchanging fluid is almost exclusively water or water steam (depending on process temperature). One of the fluids is usually passed through pipes or tubes, and the other fluid is passed round or across these. At any point in the equipment, the local temperature differences and the heat transfer coefficients control the rate of heat exchange. The fluids can flow in the same direction through the equipment (parallel flow) or in opposite directions (counter flow), or they can also flow at right angles to each other (cross flow). Various combinations of these directions of flow can occur in different parts of the exchanger; in fact, most heat exchangers of this type have a mixed flow pattern. In parallel flow, the maximal temperature difference between the coldest and the hottest stream is at the entry to the heat exchanger, but at the exit the two streams approach each other's temperature. In a counter flow exchanger, leaving streams can approach the temperatures of the entering stream of the other component and so counter flow exchangers are often preferred. To further improve heat exchange efficiency, the surface of the tubes can be "corrugated", to extend the available surface and also to provoke dynamic turbulence in the fluids, improving thermal exchange up to 90% (especially in the case of low viscosity fluids). These equipment can in principle be used also to cool down beverages, e.g. using sodium chlorine brine as exchanger fluid, but other approaches are more often used for this purpose.

## Plate Heat Exchangers

Another popular heat exchanger for fluids of low viscosity, such as milk, is the plate heat exchanger, where heating and cooling fluids flow through alternate tortuous passages between vertical plates The plates are clamped together, separated by spacing gaskets, and the heating and cooling fluids are arranged so that they flow between alternate plates. Suitable gaskets and channels control the flow and allow parallel or counter current flow in any desired number of passes. A substantial advantage of this type of heat exchanger is that it offers a large transfer surface that is readily accessible for cleaning in fact the banks of plates are usually arranged so that they may be taken apart easily. Overall heat transfer coefficients are high, in the order of 2400-6000 $J\ m^{-2}\ s^{-1}\ ^\circ C^{-1}$.

## Jacketed Pans

In this kind of heat exchanger, the fluid (liquid food up to paste consistence food can be treated with this equipment) to be heated is contained in a vessel, which may also be provided with an agitator to keep the fluid moving across the heat transfer surface, to assure its homogeneous heating. Where there is no agitation, heat transfer coefficients are lower or even halved. The source of heat is commonly steam condensing in the vessel jacket: there must be the minimum of air within the steam in the jacket, because air hinders heat exchange. The pan itself can be made of cast iron, stainless steel, or copper. Heat transfer coefficients are not very high: depending on the pan material and on the viscosity of the fluid to be heated, can range from 300 $J\ m^{-2}\ s^{-1}\ ^\circ C^{-1}$ (for pastes heated in stainless steel pans) to 1800 $J\ m^{-2}\ s^{-1}\ ^\circ C^{-1}$ (for thin liquids heated in copper pans). Furthermore, usually this equipment only allows discontinuous food processing.

## Heating Coils Immersed in Liquids

In some food processes, where quick heating is required, a helical coil may be fitted inside the pan and steam admitted to the coil. This can give greater heat transfer rates than jacketed pans, because there can be a bigger heat transfer surface and also the heat transfer coefficients are higher for coils than for the pan walls. Examples of the overall heat transfer coefficient are quoted as: 300-1400 $J\ m^{-2}\ s^{-1}\ ^\circ C^{-1}$ for sugar and molasses solutions heated with steam using a copper coil, 1800 for milk in a coil heated with water, and 3600 for a boiling aqueous solution heated with steam in the coil.

## Scraped Surface Heat Exchangers

Another kind of heat exchanger consists of a jacketed cylinder with an internal cylinder concentric to the first one, and fitted with scraping blades (or shafts), the blades rotate, causing the fluid to flow through the annular space between the cylinders with the outer heat transfer surface constantly scraped: the scraping shafts continuously remove the food from the walls, keeping it mixed and allowing optimal heat exchange. This equipment finds considerable use particularly for products of higher viscosity, and can be also used to drive away heat from the food instead of administering it (e.g. freezing of ice creams and cooling of fats during margarine manufacture). Scraped surface exchangers can also be used to process foods sensitive to heat and/or to mechanical stress. Coefficients of heat transfer vary with speeds of rotation and with heated fluid characteristics, and are quite high, in the order of 900-4000 J m$^{-2}$ s$^{-1}$ °C$^{-1}$.

# Blanching

Blanching is used to destroy enzymatic activity in vegetable and some fruits prior to other processing like freezing or dehydration or canning or thermal processing. It is a pretreatment by mild heat for a specific time followed by rapid cooling or passing immediately to the next processing stage. The time and temperature combination varies from product to product, the condition and size of product. Generally the temperature varies from 88 to 99 °C. In some of the fruits and vegetables poly phenol oxydase enzyme is responsible for discoloration in presence of oxygen, hence it needs to be inactivated by blanching pretreatment, before futher processing of fruits and vegetables to maintain its original colour after processing.

## Mechanism of Blanching

Plant cells are discrete membrane-bound structures contained within semirigid cell walls. The outer or cytoplasmic membrane acts as a skin, maintaining turgor pressure within the cell. Loss of turgor pressure leads to softening of the tissue. Within the cell are a number of organelles, including the nucleus, vacuole, chloroplasts, chromoplasts and mitochondria. This compartmentalisation is essential to the various biochemical and physical functions. Blanching causes cell death and physical and metabolic chaos within the cells. The heating effect leads to enzyme destruction as well as damage to the cytoplasmic and other membranes, which become permeable to water and solutes. An immediate effect is the loss of turgor pressure. Water and solutes may pass into and out of the cells, a major consequence being nutrient loss from the tissue. Also cell constituents, which had previously been compartmentalized in sub cellular organelles, become free to move and interact within the cell.

The following f actors are affecting blanching time:

1. The type of fruit or vegetable.

2. The size of the pieces of food.

3. The blanching temperature and

4. The method of heating.

## Purpose and Objective of Blanching

The purpose of blanching is to achieve several objectives:

1. To soften the tissue to facilitate packaging.

2. To avoid damage to the product.

3. To eliminate air form the product.

4. To preserve the natural colour.

5. To destroy or retard certain undesirable enzymes.

6. To help preserve natural flavour.

The major purpose of blanching is frequently to inactivate enzymes, which would otherwise lead to quality reduction in the processed product. For example, with frozen foods, deterioration could take place during any delay prior to processing, during freezing, during frozen storage or during subsequent thawing. Similar considerations apply to the processing, storage and rehydration of dehydrated foods. Enzyme inactivation prior to heat sterilization is less important as the severe processing will destroy any enzyme activity, but there may be an appreciable time before the food is heated to sufficient temperature, so quality may be better maintained if enzymes are destroyed prior to heat sterilisation processes such as canning.

It is important to inactivate quality-changing enzymes, that is enzymes which will give rise to loss of colour or texture, production of off odours and flavours or breakdown of nutrients. Many such enzymes have been studied, including a range of peroxidases, catalases and lipoxygenases. Peroxidase and to a lesser extent catalase are frequently used as indicator enzymes to determine the effectiveness of blanching. Although other enzymes may be more important in terms of their quality-changing effect, peroxidase is chosen because it is extremely easy to measure and it is the most heat resistant of the enzymes in question. More recent work indicates that complete inactivation of peroxidase may not be necessary and retention of a small percentage of the enzyme following blanching of some vegetables may be acceptable.

Blanching causes the removal of gases from plant tissues, especially intercellular gas. This is especially useful prior to canning where blanching helps achieve vacuum in the containers, preventing expansion of air during processing and hence reducing strain on the containers and the risk of misshapen cans and/or faulty seams. In addition, removing oxygen is useful in avoiding oxidation of the product and corrosion of the can. Removal of gases, along with the removal of surface dust, has a further effect in brightening the colour of some products, especially green vegetables.

Shrinking and softening of the tissue is a further consequence of blanching. This is of benefit in terms of achieving filled weight into containers, so for example it may be possible to reduce the tin plate requirement in canning. It may also facilitate the filling of containers. It is important to control the time/temperature conditions to avoid overprocessing, leading to excessive loss of texture in some processed products. Calcium chloride addition to blanching water helps to maintain the texture of plant tissue through the formation of calcium pectate complexes. Some weight loss from the tissue is inevitable as both water and solutes are lost from the cells.

A further benefit is that blanching acts as a final cleaning and decontamination process. It also removes pesticide residues or radionuclides from the surface of vegetables, while toxic constituents naturally present (such as nitrites, nitrates and oxalate) are reduced by leaching. Very significant reductions in microorganism content can be achieved, which is useful in frozen or dried foods where surviving organisms can multiply on thawing or rehydration. It is also useful before heat sterilization if large numbers of microorganisms are present before processing.

## Principles of Blanching

Blanching is achieved in hot water for a short period of time or in an atmosphere of steam. In water blanching, the product is moved through water usually maintained at a temperature between 88 and 99 °C. In steam blanching the product is carried on a belt through a steam chamber into which live steam is constantly injected. The steam chamber is hooded and equipped with exhaust and also a drain for the condensate. The time temperatures are regulated for each specific product to achieve the desired enzyme inactivation, colour preservation and other characteristics. As a guide, the operator utilizes either the catalase or the peroxidase tests to determine the adequacy of blanching. Currently the peroxide test is commonly used in industry. For the most part, a negative peroxidase test is necessary to prevent the development of undesirable characteristics in the finished product. Immediately after blanching, vegetables are quickly cooled, usually in cold water, which often serves as means to convey the product to the next operation. A rod type cylindrical reel connected to the discharge of blancher and equipped with water sprays also serves as an excellent cooling system.

## Processing Conditions for Blanching

It is essential to control the processing conditions accurately to avoid loss of texture, weight, colour and nutrients. All water-soluble materials, including minerals, sugars, proteins and vitamins, can leach out of the tissue,leading to nutrient loss. In addition, some nutrient loss (especially ascorbic acid) occurs through thermal lability and, to a lesser extent, oxidation.

Ascorbic acid is the most commonly measured nutrient with respect to blanching, as it covers all eventualities, being water soluble and hence prone to leaching from cells, thermally labile, as well as being subject to enzymic breakdown by ascorbic acid oxidase during storage. Wide ranges of vitamin C breakdown are observed, depending on the raw material and the method and precise conditions of processing.

The aim is to minimize leaching and thermal breakdown while completely eliminating ascorbic acid oxidase activity, such that vitamin C losses in the product are restricted to a few percent. Generally steam blanching systems give rise to lower losses of nutrients than immersion systems, presumably because leaching effects are less important.

Blanching is an example of unsteady state heat transfer involving convective heat transfer from the blanching medium and conduction within the food piece. Mass transfer of material into and out of the tissue is also important. The precise blanching conditions (time and temperature) must be evaluated for the raw material and usually represent a balance between retaining the quality characteristics of the raw material and avoiding over-processing.

The following factors must be considered for deciding processing conditions of blanching:

1. Fruit or vegetable properties, especially thermal conductivity, which will be determined by type, cultivar, degree of maturity, etc.

2. Overall blanching effect required for the processed product, which could be expressed in many ways including: achieving a specified central temperature, achieving a specified level of peroxidase inactivation, retaining a specified proportion of vitamin C.

3. Size and shape of food pieces.

4. Method of heating and temperature of blanching medium.

Time/temperature combinations vary very widely for different foods and different processes and must be determined specifically for any situation. Holding times of 1–15 minutes at 70–100 °C are normal.

## Methods of Blanching

The two most widespread commercial methods of blanching involve passing food through an atmosphere of saturated steam or a bath of hot water. Both types of equipment are relatively simple and inexpensive. Microwave blanching is not yet used commercially on a large scale. There have been substantial developments to blanchers in recent years to reduce the energy consumption and also to reduce the loss of soluble components of foods, which reduces the volume and polluting potential of effluents and increases the yield of product.

Conventional steam blanching consists of conveying the material through an atmosphere of steam in a tunnel on a mesh belt. Uniformity of heating is often poor where food is unevenly distributed; and the cleaning effect on the food is limited.

However, the volumes of waste water are much lower than for water blanching. Fluidised bed designs and 'individual quick blanching' (a three-stage process in which vegetable pieces are heated rapidly in thin layers by steam), held in a deep bed to allow temperature equilibration, (followed by cooling in chilled air) may overcome the problems of nonuniform heating and lead to more efficient systems.

The two main conventional designs of hot water blancher are *reel* and *pipe* designs. In reel blanchers, the food enters a slowly rotating mesh drum which is partly submerged in hot water. The heating time is determined by the speed of rotation. In pipe blanchers, the food is in contact with hot water recirculating through a pipe. The residence time is determined by the length of the pipe and the velocity of the water. There is much scope for improving energy efficiency and recycling water in either steam or hot water systems. Blanching may be combined with peeling and cleaning operations to reduce costs.

Following the microwave heating, the vegetable material is subjected to blanching comprising heat treating in a current of hot air at temperature 100 to 150°C. The heating is conducted in an environment which prevents loss of water from the vegetable material and this may readily achieved by introducing steam in to the oven.

## Equipment for Blanching

### Steam Blanchers

At its simplest a steam blancher consists of a mesh conveyor belt that carries food through a steam atmosphere. The residence time of the food is controlled by the speed of the conveyor. In conventional steam blanching, there is often poor uniformity of heating in the multiple layers of food. To overcome this Individual Quick Blanching (IQB) was introduced which involves blanching in two stages. In the first stage food is heated in single layer to a sufficiently high temperature. In the second stage a deep bed of food is held for sufficient time to allow the temperature at the center of each piece to increase to that needed for enzyme inactivation.

## Hot Water Blanchers

There are a number of different designs of blanchers each of which retains the food in hot water at 70 – 100 °C for a specific time and thus removes it to a dewatering-cooling section. A blancher cooler is shown in the figure. It has three sections: a pre-heating stage, a launching stage and a cooling stage. The food is preheated with water that is circulated through a heat exchanger. After blanching a second re-circulation system cools the food. The two systems pass water through the heat exchanger and this heats the pre-heat water and simultaneously cools the cooling water. A re-circulated water-steam mixture is used to blanch the food and final cooling is done by cold air.

# Sous Vide Cooking

Sous vide cooking encompasses two types of products that differ in their production patterns – cook-hold or cook-serve and cook-chill or cook-freeze food. Cook-hold or cook-serve sous vide techniques are intended for uninterrupted food processing in restaurants and households and include the following steps: food preparation, vacuum packaging, heating or pasteurising, finishing and serving. On the other hand, cook-chill or cook-freeze sous vide techniques include food preparation, vacuum packaging, heating or pasteurising, and finally refrigerating or rapid freezing in industrial conditions. This ready-to-take type of sous vide product is intended for use at home after reheating or re-thermalising and serving.

In contrast to conventional cooking, the temperatures of sous vide processing are typically in the range from 65 to 95°C, with temperatures commonly exceeding 70°C, but not 100°C. There are four general time-temperature regimes for sous vide processing: i) 90°C for 10 minutes; ii) 70°C for 2 minutes (this process results in a large log reduction of vegetative cells, without affecting spores, and the products obtained by application of this temperature-time regime are usually frozen after processing); iii) minimum heating process with optional pasteurisation; and iv) light processing, which refers to neither cooking nor pasteurisation.

The lowest sous vide temperatures are applied to fish, seafood and meat processing (50–75°C, with an average temperature of 55°C) and maintained for several hours or even days, while the highest temperatures of 90–100°C (with an average temperature of 85°C) are applied to vegetable processing, which typically takes only a few minutes. For red meat which is cooked for less than 4 hours, the average temperature is 56°C, while for red meat which is cooked for 4 hours or more, the average temperature is up to 60°C. The average sous vide temperature for poultry is 63.5°C, egg products are prepared at 64.5°C and dairy products at 82°C.

The shelf-life of sous vide products depends on both the temperature-time treatment and the storage temperature, and typically ranges from 6 to 42 days.

# Pasteurisation

Pasteurisation is a process that slows down microbial growth in food. The aim of pasteurization is not to completely destroy all pathogenic micro-organisms in foods (typically in milk & milk products); but just to reduce the number of viable pathogens so that they are unable to cause disease if the pasteurized product is stored as indicated and consumed before its expiry date. Now-a-days a new method of pasteurisation known as flash pasteurisation (HTST) is used which exposes foods and liquids to short periods of high temperature.

In the same way pasteurization of milk is used to get rid of disease producing bacteria and reduce the total bacteria count in milk substantially. This helps to maintain quality of milk. For milk pasteurisation the current trend is to use low temperature –long-time (LTLT) method. In this method the temperature is held at 63°C for 30 min to eliminate pathogenic bacteria that may be present such as Mycobacterium tuberculosis and Coxiella burnett. On the other hand milk is heated to 72°C for 15 sec in high temperature short time pasteurisation (HTST).

In ultrahigh temperature (UHT) pasteurisation milk and milk products are heated to at least 138°C for 2 sec and packaged aseptically. Pasteurised milk is not sterile it must be quickly cooled after pasteurisation to prevent multiplication of surviving bacteria. The effectiveness of pasteurisation is evaluated by phosphatase test which is alkaline phosphatase activity in milk.

Food Safety Standards Regulation says that the terms "Pasteurisation", "Pasteurised" and similar terms shall be taken to refer to the process of heating every particle of milk of different classes to at least to 63°C for 30 min or heating it to at least 71.5°C and holding it at that temperature for 15 seconds or any other approved temperature time combination that will serve to give a negative Phosphatase Test. All pasteurised milk of different classes shall be cooled immediately to a temperature of 10°C or less.

The regulations also say that recombined milk that is a homogenised product prepared from milk fat, non-fat-milk solids and water shall be pasteurised and shall show a negative Phosphatase test. The regulations also say that no person shall either by himself or by any servant or agent sell—dahi or curd not prepared from boiled, pasteurised or sterilized milk.

# Heat Sterilization

The time and temperature required for the sterilization of foods are influenced by several factors, including the type of microorganisms found on the food, the size of the container, the acidity or pH of the food, and the method of heating.

The thermal processes of canning are generally designed to destroy the spores of the bacterium C. botulinum. This microorganism can easily grow under anaerobic conditions, producing the deadly toxin that causes botulism. Sterilization requires heating to temperatures greater than 100 °C (212 °F). However, C. botulinum is not viable in acidic foods that have a pH less than 4.6. These foods can be adequately processed by immersion in water at temperatures just below 100 °C.

The sterilization of low-acid foods (pH greater than 4.6) is generally carried out in steam vessels called retorts at temperatures ranging from 116 to 129 °C (240 to 265 °F). The retorts are controlled by automatic devices, and detailed records are kept of the time and temperature treatments for each lot of processed cans. At the end of the heating cycle, the cans are cooled under water sprays or in water baths to approximately 38 °C (100 °F) and dried to prevent any surface rusting. The cans are then labeled, placed in fibreboard cases either by hand or machine, and stored in cool, dry warehouses.

Sterilization is necessary for the complete destruction or removal of all microorganisms (including spore-forming and non-sporeforming bacteria, viruses, fungi, and protozoa) that could contaminate pharmaceuticals or other materials and thereby constitute a health hazard. Since the achievement of the absolute state of sterility cannot be demonstrated, the sterility of a pharmaceutical preparation can be defined only in terms of probability. The efficacy of any sterilization process will depend on the nature of the product, the extent and type of any contamination, and the conditions under which the final product has been prepared. The requirements for Good Manufacturing Practice should be observed throughout all stages of manufacture and sterilization.

Classical sterilization techniques using saturated steam under pressure or hot air are the most reliable and should be used whenever possible. Other sterilization methods include filtration, ionizing radiation (gamma and electron-beam radiation), and gas (ethylene oxide, formaldehyde).

For products that cannot be sterilized in the final containers, aseptic processing is necessary. Materials and products that have been sterilized by one of the above processes are transferred to presterilized containers and sealed, both operations being carried out under controlled aseptic conditions.

Whatever method of sterilization is chosen, the procedure must be validated for each type of product or material, both with respect to the assurance of sterility and to ensure that no adverse change has taken place within the product. Failure to follow precisely a defined, validated process could result in a non-sterile or deteriorated product. A typical validation programme for steam or dry-heat sterilization requires the correlation of temperature measurements, made with sensory devices to demonstrate heat penetration and heat distribution, with the destruction of biological indicators, i.e. preparations of specific microorganisms known to have high resistance to the particular sterilization process. Biological indicators are also used to validate other sterilization methods, and sometimes for routine control of individual cycles. Periodic revalidation is recommended.

## Heating in an Autoclave (Steam Sterilization)

Exposure of microorganisms to saturated steam under pressure in an autoclave achieves their destruction by the irreversible denaturation of enzymes and structural proteins. The temperature at which denaturation occurs varies inversely with the amount of water present. Sterilization in saturated steam thus requires precise control of time, temperature, and pressure. As displacement of the air by steam is unlikely to be readily achieved, the air should be evacuated from the autoclave before admission of steam. This method should be used whenever possible for aqueous preparations and for surgical dressings and medical devices.

The recommendations for sterilization in an autoclave are 15 minutes at 121-124 °C (200 kPa). The temperature should be used to control and monitor the process; the pressure is mainly used to obtain the required steam temperature. Alternative conditions, with different combinations of time and temperature, are given below.

1 atm = 101 325 Pa

| Temperature (°C) | Approximate corresponding pressure (kPa) | Minimum sterilization time (min) |
|---|---|---|
| 126-129 | 250 (~2.5 atm) | 10 |
| 134-138 | 300 (~3.0 atm) | 5 |

Minimum sterilization time should be measured from the moment when all the materials to be sterilized have reached the required temperature throughout. Monitoring the physical conditions within the autoclave during sterilization is essential. To provide the required information, temperature-monitoring probes should be inserted into representative containers, with additional probes placed in the load at the potentially coolest parts of the loaded chamber (as established in the course of the validation programme). The conditions should be within ± 2 °C and ± 10 kPa (± 0.1 atm) of the required values. Each cycle should be recorded on a time-temperature chart or by other suitable means.

Aqueous solutions in glass containers usually reach thermal equilibrium within 10 minutes for volumes up to 100 mL and 20 minutes for volumes up to 1000 mL.

Porous loads, such as surgical dressings and related products, should be processed in an apparatus that ensures steam penetration. Most dressings are adequately sterilized by maintaining them at a temperature of 134 - 138 °C for 5 minutes.

In certain cases, glass, porcelain, or metal articles are sterilized at 121 - 124 °C for 20 minutes.

Fats and oils may be sterilized at 121 °C for 2 hours but, whenever possible, should be sterilized by dry heat.

In certain cases (e.g. thermolabile substances), sterilization may be carried out at temperatures below 121 °C, provided that the chosen combination of time and temperature has been validated. Lower temperatures offer a different level of sterilization; if this is evaluated in combination with the known microbial burden of the material before sterilization, the lower temperatures may be satisfactory. Specific conditions of temperature and time for certain preparations are stated in individual monographs.

The bioindicator strain proposed for validation of this sterilization process is: spores of Bacillus stearothermophilus (e.g. ATCC 7953 or CIP 52.81) for which the D-value (i.e. 90% reduction of the microbial population) is 1.5-2 minutes at 121 °C, using about $10^6$ spores per indicator.

## Dry-heat Sterilization

In dry-heat processes, the primary lethal process is considered to be oxidation of cell constituents. Dry-heat sterilization requires a higher temperature than moist heat and a longer exposure time. The method is, therefore, more convenient for heat-stable, non-aqueous materials that cannot be sterilized by steam because of its deleterious effects or failure to penetrate. Such materials include glassware, powders, oils, and some oil-based injectables.

Preparations to be sterilized by dry heat are filled in units that are either sealed or temporarily closed for sterilization. The entire content of each container is maintained in the oven for the time and at the temperature given in the table below. Other conditions may be necessary for different preparations to ensure the effective elimination of all undesirable microorganisms.

| Temperature (°C) | Minimum sterilization time (min) |
|:---:|:---:|
| 160 | 180 |

| 170 | 60 |
|-----|-----|
| 180 | 30 |

Specific conditions of temperature and time for certain preparations are stated in individual monographs.

The oven should normally be equipped with a forced air system to ensure even distribution of heat throughout all the materials processed. This should be controlled by monitoring the temperature. Containers that have been temporarily closed during the sterilization procedure are sealed after sterilization using aseptic techniques to prevent microbial recontamination.

The bioindicator strain proposed for validation of the sterilization process is: spores of Bacillus subtilis (e.g. var. niger ATCC 9372 or CIP 77.18) for which the D-value is 5-10 minutes at 160 °C using about $10^6$ spores per indicator.

## Filtration

Sterilization by filtration is employed mainly for thermolabile solutions. These may be sterilized by passage through sterile bacteria-retaining filters, e.g. membrane filters (cellulose derivatives, etc.), plastic, porous ceramic, or suitable sintered glass filters, or combinations of these. Asbestos-containing filters should not be used.

Appropriate measures should be taken to avoid loss of solute by adsorption onto the filter and to prevent the release of contaminants from the filter. Suitable filters will prevent the passage of microorganisms, but the filtration must be followed by an aseptic transfer of the sterilized solution to the final containers which are then immediately sealed with great care to exclude any recontamination.

Usually, membranes of not greater than 0.22 µm nominal pore size should be used. The effectiveness of the filtration method must be validated if larger pore sizes are employed.

To confirm the integrity of filters, both before and after filtration, a bubble point or similar test should be used, in accordance with the filter manufacturer's instructions. This test employs a prescribed pressure to force air bubbles through the intact membrane previously wetted with the product, with water, or with a hydrocarbon liquid.

All filters, tubes, and equipment used "downstream" must be sterile. Filters capable of withstanding heat may be sterilized in the assembly before use by autoclaving at 121 °C for 15 - 45 minutes depending on the size of the filter assembly. The effectiveness of this sterilization should be validated. For filtration of a liquid in which microbial growth is possible, the same filter should not be used for procedures lasting longer than one working day.

## Exposure to Ionizing Radiation

Sterilization of certain active ingredients, drug products, and medical devices in their final container or package may be achieved by exposure to ionizing radiation in the form of gamma radiation from a suitable radioisotopic source such as $^{60}$Co (cobalt 60) or of electrons energized by a suitable electron accelerator. Laws and regulations for protection against radiation must be respected.

Gamma radiation and electron beams are used to effect ionization of the molecules in organisms. Mutations are thus formed in the DNA and these reactions alter replication. These processes are very dangerous and only well-trained and experienced staff should decide upon the desirability of their use and should ensure monitoring of the processes. Specially designed and purpose built installations and equipment must be used.

It is usual to select an absorbed radiation level of 25 kGy (2.5 Mrad), although other levels may be employed provided that they have been validated.

- Kilogray
- Megarad

Radiation doses should be monitored with specific dosimeters during the entire process. Dosimeters should be calibrated against a standard source on receipt from the supplier and at appropriate intervals thereafter. The radiation system should be reviewed and validated whenever the source material is changed and, in any case, at least once a year.

The bioindicator strains proposed for validation of this sterilization process are: spores of Bacillus pumilus (e.g. ATCC 27142 or CIP 77.25) with 25 kGy (2.5 Mrad) for which the D-value is about 3 kGy (0.3 Mrad) using $10^7$ -$10^8$ spores per indicator; for higher doses, spores of Bacillus cereus (e.g. SSI C 1/1) or Bacillus sphaericus (e.g. SSl C$_1$ A) are used.

## Gas Sterilization

The active agent of the gas sterilization process can be ethylene oxide or another highly volatile substance. The highly flammable and potentially explosive nature of such agents is a disadvantage unless they are mixed with suitable inert gases to reduce their highly toxic properties and the possibility of toxic residues remaining in treated materials. The whole process is difficult to control and should only be considered if no other sterilization procedure can be used. It must only be carried out under the supervision of highly skilled staff.

The sterilizing efficiency of ethylene oxide depends on the concentration of the gas, the humidity, the time of exposure, the temperature, and the nature of the load. In particular, it is necessary to ensure that the nature of the packaging is such that the gas exchange can take place. It is also important to maintain sufficient humidity during sterilization. Records of gas concentration and of temperature and humidity should be made for each cycle. Appropriate sterilization conditions must be determined experimentally for each type of load.

After sterilization, time should be allowed for the elimination of residual sterilizing agents and other volatile residues, which should be confirmed by specific tests.

Because of the difficulty of controlling the process, efficiency must be monitored each time using the proposed bioindicator strains: spores of Bacillus subtilis (e.g. var. niger ATCC 9372 or CIP 77.18) or of Bacillus stearothermophilus, (e.g. ATCC 7953 or CIP 52.81). The same quantity of spores should be used as for "Heating in an autoclave" and "Dry-heat sterilization".

# Evaporation

Most food liquids have relatively low solids contents. For example, whole milk contains approximately 12.5% total solids, fruit juice 12%, sugar solution after extraction from sugar beet 15%, solution of coffee solutes after extraction from ground roasted beans 25%. The most common method used to achieve this is to "boil off" or evaporate some of the water by the application of heat. Other methods used to concentrate food liquids are freeze concentration and membrane separation. If evaporation is carried out in open pans at atmospheric pressure, the initial temperature at which the solution boils will be some degrees above 100C, depending on the solids content of the liquid. As the solution becomes more concentrated, the evaporation temperature will rise. It could take from several minutes to a few hours to attain the solids content required. Exposure of the food liquid to these high temperatures for these lengths of time is likely to cause changes in the colour and flavour of the liquid. In some cases such changes may be acceptable, or even desirable, for example when concentrating sugar solutions for toffee manufacture or when reducing gravies. However, in the case of heat-sensitive liquids such as milk or fruit juice, such changes are undesirable. To reduce such heat damage, the pressure above the liquid in the evaporator may be reduced below atmospheric by means of condensers, vacuum pumps or steam ejectors. Since a liquid boils when the vapour pressure it exerts equals the external pressure above it, reducing the pressure in the evaporator lowers the temperature at which the liquid will evaporate. Typically, the pressure in the evaporator will be in the range 7.5– 85.0 kPa absolute, corresponding to evaporation temperatures in the range 40– 95 °C. The use of lower pressures is usually uneconomic. This is known as vacuum evaporation. The relatively low evaporation temperatures which prevail in vacuum evaporation mean that reasonable temperature differences can be maintained between the heating medium, saturated steam, and the boiling liquid, while using relatively low steam pressures. This limits undesirable changes in the colour and flavour of the product. For aqueous liquids, the relationship between pressure and evaporation temperature may be obtained from thermodynamic tables and psychrometric charts. Relationships are available for estimating the evaporating temperatures of nonaqueous liquids at different pressures.

Another factor which affects the evaporation temperature, is known as the boiling point rise (BPR) or boiling point elevation (BPE). The boiling point of a solution is higher than that of the pure solvent at the same pressure. The higher the soluble solids content of the solution, the higher its boiling point. Thus, the initial evaporation temperature will be some degrees above that corresponding to pressure in the evaporator, depending on the soluble solids content of the feed. However, as evaporation proceeds and the concentration of the soluble solids increases, the evaporation temperature rises. This is likely to result in an increase in changes in the colour and flavour of the product. If the temperature of the steam used to heat the liquid is kept constant, the temperature difference between it and the evaporating liquid decreases. This reduces the rate of heat transfer and hence the rate of evaporation. To maintain a constant rate of evaporation, the steam pressure may be increased. However, this is likely to result in a further decrease in the quality of the product. Data on the BPR in simple solutions and some more complex foods is available in the literature in the form of plots and tables. Relationships for estimating the BPR with increase in solids concentrations have also been proposed. BPR may range from < 1 °C to 10 °C in food liquids. For example, the BPR of a sugar solution containing 50% solids is about 7 °C.

In some long tube evaporators, the evaporation temperature increases with increase in the depth of the liquid in the tubes in the evaporator, due to hydrostatic pressure. This can lead to overheating of the liquid and heat damage. This factor has to be taken into account in the design of evaporators and in selecting the operating conditions, in particular, the pressure of the steam in the heating jacket.

The viscosity of most liquids increases as the solids content increases during evaporation. This can lead to a reduction in the circulation rates and hence the rates of heat transfer in the heating section of the evaporator. This can influence the selection of the type of evaporator for a particular liquid food. Falling film evaporators are often used for moderately viscous liquids. For very viscous liquids, agitated thin film evaporators are used. Thixotropic (or time-dependent) liquids, such as concentrated tomato juice, can pose special problems during evaporation. The increase in viscosity can also limit the maximum concentration attainable in a given liquid.

Fouling of the heat transfer surfaces may occur in evaporators. This can result in a decrease in the rate of heat transfer and hence the rate of evaporation. It can also necessitate expensive cleaning procedures. Fouling must be taken into account in the design of evaporators and in the selection of the type of evaporator for a given duty. Evaporators that feature forced circulation of the liquid or agitated thin films are used for liquids that are susceptible to fouling.

Some liquids are prone to foaming when vigorously boiling in an evaporator. Liquids which contain surface active foaming agents, such as the proteins in skimmed milk, are liable to foam. This can reduce rates of heat transfer and hence rates of evaporation. It may also result in excessive loss of product by entrainment in the vapour leaving the heating section. This in turn can cause contamination of the cooling water to spray condensers and lead to problems in the disposal of that effluent. In some cases, anti-foaming agents may be added to the feed to reduce foaming. Care must be taken not to infringe any regulations by the addition of such aids.

Volatile aroma and flavour compounds may be lost during vacuum evaporation, resulting in a reduction in the organoleptic quality of products such as fruit juices or coffee extract. In the case of fruit juices, this loss may be partly offset by adding some of the original juice, known as "cut back juice", to the concentrate. Alternatively, the volatiles may be stripped from the vapour, concentrated and added to the concentrated liquid.

## Equipment used in Vacuum Evaporation

A single-effect vacuum evaporator has the following components.

A heat exchanger, known as a calandria, by means of which the necessary sensible and latent heat is supplied to the feed to bring about the evaporation of some of the liquid. Saturated steam is the usual heating medium but hot water and other thermal fluids are sometimes used. Tubular and plate exchangers of various designs are widely used. Other, more sophisticated designs are available, including agitated thin film models, expanding flow chambers and centrifugal exchangers.

A device to separate the vapour from the concentrated liquid phase. In vacuum evaporators, mechanical devices such as chambers fitted with baffles or meshes and cyclone separators are used to reduce entrainment losses.

A condenser to convert the vapour back to a liquid and a pump, steam ejector or barometric leg to remove the condensate, thus creating and maintaining the partial vacuum in the system.

Most evaporators are constructed in stainless steel except where there are extreme corrosion problems.

The following types of evaporators are used in the food industry.

## Vacuum Pans

A hemispherical pan equipped with a steam jacket and sealed lid, connected to a vacuum system, is the simplest type of vacuum evaporator in use in industry. The heat transfer area per unit volume is small and so the time required to reach the desired solids content can run into hours. Heating occurs by natural convection.

However, an impeller stirrer may be introduced to increase circulation and reduce fouling. Small pans have a more favourable heat transfer area to volume ratio. They are useful for frequent changes of product and for low or variable throughputs. They are used in jam manufacture, the preparation of sauces, soups and gravies and in tomato pulp concentration.

## Short Tube Vacuum Evaporators

This type of evaporator consists of a calandria made up of a bundle of short vertical tubes surrounded by a steam jacket, located near the bottom of a large vessel. The tubes are typically 25–75 mm in diameter and 0.5–2.0 m long. The liquid being concentrated normally covers the calandria. Steam condensing on the outside of the tubes heats the liquid causing it to rise by natural convection. Some of the water evaporates and flows to the condenser. The liquid circulates down through the larger, cooler tube in the centre of the bundle, known as the downcomer. This type of evaporator is suitable for low to moderate viscosity liquids, which are not very heat-sensitive. With viscous liquids heat transfer rates are low, hence residence times are relatively long and there is a high risk of fouling. Sugar solutions, glucose and malt extract are examples of products concentrated in this type of evaporator. It can also be used for crystallisation operations. For this application an impeller may be located in the downcomer to keep the crystals in suspension.

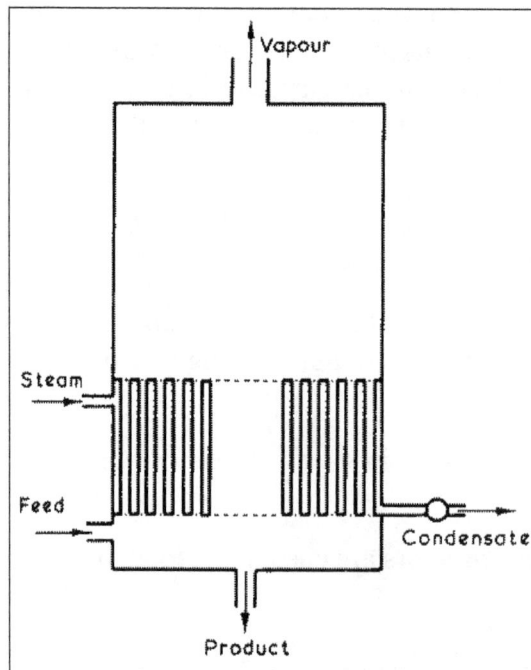

Vertical short tube evaporator.

In another design of a short tube evaporator the calandria is external to separator chamber and may be at an angle to the vertical. The liquid circulates by natural convection within the heat exchanger and also through the separation chamber. The liquid enters the separation chamber tangentially. A swirling flow pattern develops in the chamber, generating centrifugal force, which assists in separating the vapour from the liquid. The vigorous circulation of the liquid results in relatively high rates of heat transfer. It also helps to break up any foam which forms. The tubes are easily accessible for cleaning. Such evaporators are also used for concentrating sugar solutions, glucose and malt and more heat-sensitive liquids such as milk, fruit juices and meat extracts.

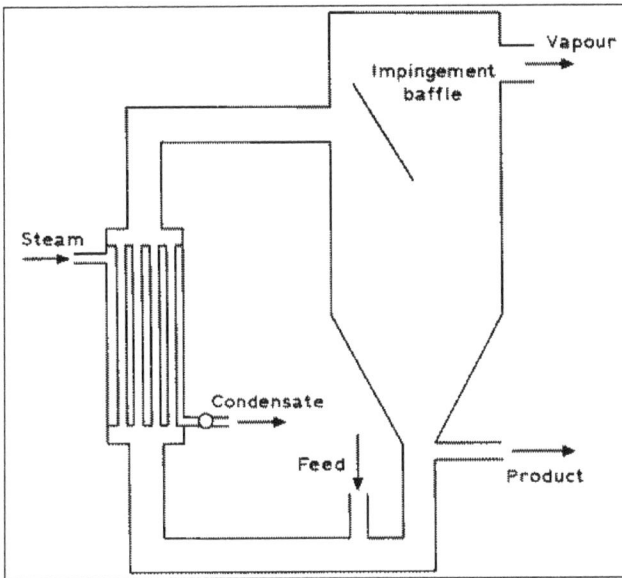

Natural circulation evaporator.

A pump may be introduced to assist in circulating more viscous liquids. This is known as forced circulation. The choice of pump will depend on the viscosity of the liquid. Centrifugal pumps are used for moderately viscous materials, while positive displacement pumps are used for very viscous liquids.

## Long Tube Evaporators

These consist of bundles of long tubes, 3–15 m long and 25–50 mm in diameter, contained within a vertical shell into which steam is introduced. The steam condensing on the outside of the tubes provides the heat of evaporation. There are three patterns of flow of the liquid through such evaporators.

In the climbing film evaporator the preheated feed is introduced into the bottom of the tubes. Evaporation commences near the base of the tubes. As the vapour expands, ideally, it causes a thin film of liquid to rise rapidly up the inner walls of the tubes around a central core of vapour. In practice, slugs of liquid and vapour bubbles also rise up the tubes. The liquid becomes more concentrated as it rises. At the top, the liquid-vapour mixture enters a cyclone separator. The vapour is drawn off to a condenser and pump, or into the heating jacket of another calandria, in a multiple-effect system. The concentrated liquid may be removed as product, recycled through the calandria or fed to another calandria, in a multiple-effect system. The residence time of the liquid in the tubes is relatively short. High rates of heat transfer are attainable in this type of evaporator, provided there are relatively large temperature differences between the heating medium and the liquid being concentrated. However, when these temperature differences are low, the heat transfer rates are also low. This type of evaporator is suitable for low viscosity, heat-sensitive liquids such as milk and fruit juices.

In the falling film evaporator the preheated feed is introduced at the top of the tube bundle and distributed to the tubes so that a thin film of the liquid flows down the inner surface of each tube, evaporating as it descends. Uniform distribution of the liquid so that the inner surfaces of the tubes are uniformly wetted is vital to the successful operation of this type of evaporator. From the bottom of the tubes, the liquid-vapour mixture passes into a centrifugal separator and from there the liquid and vapour streams are directed in the same manner as in the climbing film evaporator. High rates of liquid flow down the tubes are attained by a combination of gravity and the expansion of the vapour, resulting in short residence times. These evaporators are capable of operating with small temperature differences between the heating medium and the liquid and can cope with viscous materials. Consequently, they are suitable for concentrating heat-sensitive foods and are very widely used in the dairy and fruit juice processing sections of the food industry today.

A climbing-falling film evaporator is also available. The feed is first partially concentrated in a climbing film section and then finished off in a falling film section. High rates of evaporation are attainable in this type of plant.

## Plate Evaporators

In these evaporators the calandria is a plate heat exchanger, similar to that used in pasteurising and sterilising liquids. The liquid is pumped through the heat exchanger, passing on one side of an assembly of plates, while steam passes on the other side. The spacing between plates is greater than that in pasteurisers to accommodate the vapour produced during evaporation. The liquid usually follows a climbing-falling film flow pattern. However, designs featuring only a falling film flow pattern are also available. The mixture of liquid and vapour leaving the calandria passes into a cyclone separator. The vapour from the separator goes to a condenser or into the heating jacket of the next stage, in a multiple-effect system. The concentrate is collected as product or goes to another stage. The advantages of plate evaporators include: high liquid velocities leading to high rates of heat transfer, short residence times and resistance to fouling. They are compact and easily dismantled for inspection and maintenance. However, they have relatively high capital costs and low throughputs. They can be used for moderately viscous, heat-sensitive liquids such as milk, fruit juices, yeast and meat extracts.

# Agitated Thin Film Evaporators

For very viscous materials and/or materials which tend to foul, heat transfer may be increased by continually wiping the boundary layer at the heat transfer surface. An agitated thin film evaporator consists of a steam jacketed shell equipped with a centrally located, rotating shaft carrying blades which wipe the inner surface of the shell. The shell may be cylindrical and mounted either vertically or horizontally. Horizontal shells may be cone-shaped, narrowing in the direction of flow of the liquid. There may be a fixed clearance of 0.5–2.0 mm between the edge of the blades and the inner surface of the shell. Alternatively, the blades may float and swing out towards the heat transfer surface as the shaft rotates, creating a film of liquid with a thickness as little as 0.25 mm. Most of the evaporation takes place in the film that forms behind the rotating blades. Relatively high rates of heat transfer are attained and fouling and foaming are inhibited. However, these evaporators have relatively high capital costs and low throughputs. They are used as single-effect units, with relatively large temperature differences between the steam and the liquid being evaporated. They are often used as "finishers" when high solids concentrations are required. Applications include tomato paste, gelatin solutions, milk products, coffee extract and sugar products.

# Centrifugal Evaporators

In this type of evaporator a rotating stack of cones is housed in a stationary shell. The cones have steam on alternate sides to supply the heat. The liquid is fed to the undersides of the cones. It forms a thin film, which moves quickly across the surface of the cones, under the influence of centrifugal force, and rapid evaporation occurs. Very high rates of heat transfer, and so very short residence times, are attained. The vapour and concentrate are separated in the shell surrounding the cones. This type of evaporator is suitable for heat-sensitive and viscous materials. High capital costs and low throughputs are the main limitations of conical evaporators. Applications include fruit and vegetable juices and purees and extracts of coffee and tea.

Vapour-Concentrate Separators: The mixture of vapour and liquid concentrate leaving the calandria needs to be separated and entrainment of droplets of the liquid in the vapour minimised. Entrained droplets represent a loss of product. They can also reduce the energy value of the vapour which would make it less effective in multiple-effect systems or when vapour recompression is being used. Separation may be brought about by gravity. If sufficient headspace is provided above the calandria, the droplets may fall back into the liquid. Alternatively, this may occur in a second vessel, known as a flash chamber. Baffles or wire meshes may be located near the vapour outlet. Droplets of liquid impinge on these, coalesce into larger droplets and drain back into the liquid under gravity. The mixture of vapour and liquid may be directed tangentially, at high velocity, either by natural or forced circulation, into a cyclone separator. Centrifugal force is developed and the more dense liquid droplets impinge on the inner wall of the chamber, lose their kinetic energy and drain down into the liquid. This type of separator is used in most long tube, plate and agitated thin film evaporators.

Condensers and Pumps: The water vapour leaving the calandria contains some noncondensable gases, which were in the feed or leaked into the system. The water vapour is converted back to a liquid in a condenser. Condensers may be of the indirect type, in which the cooling water does not mix with the water vapour. These are usually tubular heat exchangers. They are relatively expensive. Indirect condensers are used when volatiles are being recovered from the vapour or to facilitate effluent disposal. Direct condensers, known as jet or spray condensers, are more widely used. In these a spray of water is mixed with the water vapour, to condense it. A condensate pump or barometric leg is used to remove the condensed vapour. The non-condensable gases are removed by positive displacement vacuum pumps or steam jet ejectors.

## Multiple-effect Evaporation (MEE)

In a single-effect evaporator it takes 1.1–1.3 kg of steam to evaporate 1.0 kg of water. This is known as the specific steam consumption of an evaporator. The specific steam consumption of driers used for liquid foods is much higher than this. A spray drier has a specific steam consumption of 3.0–3.5 kg of steam to evaporate 1.0 kg of water. That of a drum drier is slightly lower. Thus, it is common practice to concentrate liquid foods by vacuum evaporation before drying them in such equipment. It is also common practice to preheat the liquid to its evaporation temperature before feeding it to the evaporator. This improves the specific steam consumption. The vapour leaving a single-effect evaporator contains useful heat. This vapour may be put to other uses, for example, to heat water for cleaning. However, the most widely used method of recovering heat from the vapour leaving an evaporator is multiple-effect evaporation, the principle of which is shown in figure.

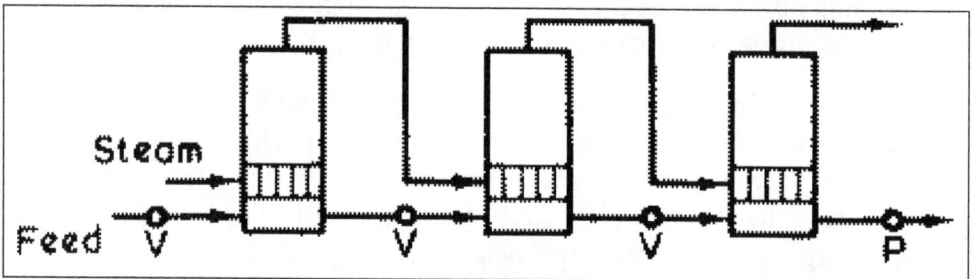

The principle of multiple-effect evaporation, with forward feeding.

The vapour leaving the separator of the first evaporator, effect 1, enters the steam jacket of effect 2, where it heats the liquid in that effect, causing further evaporation. The vapour from effect 2 is used to heat the liquid in effect 3 and so on. The vapour from the last effect goes to a condenser. The liquid also travels from one effect to the next, becoming more concentrated as it does so. This arrangement is only possible if the evaporation temperature of the liquid in effect 2 is lower than the temperature of the vapour leaving effect 1. This is achieved by operating effect 2 at a lower pressure than effect 1. This principle can be extended to a number of effects. The pressure in effect 1 may be atmospheric or slightly below. The pressure in the following effects decreases with the number of effects. The specific steam consumption in a double effect evaporator is in the range 0.55–0.70 kg of steam per 1.0 kg of water evaporated, while in a triple effect it is 0.37–0.45 kg. However, the capital cost of a multiple-effect system increases with the number of effects. The size of each effect is normally equivalent to that of a single-effect evaporator with the same capacity, working under similar operating conditions. Thus a three-effect unit will cost approximately three times that of a single effect. It must be noted that MEE does not increase the throughput above that of a single effect. Its purpose is to reduce the steam consumption. Three to five effects are most commonly used. Up to seven effects are in use in some large milk and fruit juice processing plants.

The flow pattern in figure. is known as forward feeding and is the most commonly used arrangement. Other flow patterns, including backward and mixed feeding, are also used. Each has its own advantages and limitations.

## Vapour Recompression

This procedure involves compressing some or all of the vapour from the separator of an evaporator to a pressure that enables it to be used as a heating medium. The compressed vapour is returned to the jacket of the calandria. This reduces the amount of fresh steam required and improves the specific steam consumption of the evaporator. The vapour may be compressed thermally or mechanically.

In thermal vapour recompression (TVR) the vapour from the separator is divided into two streams. One stream goes to the condenser or to the next stage of a multiple-effect system. The other enters a steam jet compressor fed with fresh high pressure steam. As it passes through the jet of the compressor the pressure of the fresh steam falls and it mixes with the vapour from the evaporator. The vapour mix then passes through a second converging-diverging nozzle where the pressure increases. This high pressure mixture then enters the jacket of the calandria.

In mechanical vapour recompression (MVR) all the vapour from the separator is compressed in a mechanical compressor which can be driven by electricity, a gas turbine or a steam turbine. The compressed vapour is then returned to the jacket of the calandria. Both methods are used in industry. MVR is best suited to large capacity duties. Both TVR and MVR are used in conjunction multipleeffect evaporators. Vapour recompression may be applied to one or more of the effects. A specific steam consumption of less than 0.10 kg steam per 1.0 kg of water evaporated is possible. A seven-effect falling film system, with TVR applied to the second effect, the compressed vapour being returned to the first effect.

## Applications for Evaporation

The purposes for which evaporation is used in the food industry include: to produce concentrated liquid products (for sale to the consumer or as ingredients to be used in the manufacture of other consumer products), to preconcentrate liquids for further processing and to reduce the cost of transport, storage and in some cases packaging, by reducing the mass and volume of the liquid.

## Concentrated Liquid Products

### Evaporated (Unsweetened Condensed) Milk

The raw material for this product is whole milk which should of good microbiological quality. The first step is standardisation of the composition of the raw milk so as to produce a finished product with the correct composition. The composition of evaporated milk is normally 8% fat and 18% solids not fat (snf) but this may vary from country to country. Skim milk and/or cream is added to the whole milk to achieve the correct fat:snf ratio. To prevent coagulation during heat processing and to minimise age thickening during storage, the milk is stabilised by the addition of salts, including phosphates, citrates and bicarbonates, to maintain pH 6.6–6.7. The milk is then heat treated. This is done to reduce the microbiological load and to improve its resistance to coagulation during subsequent sterilisation. The protein is denatured and some calcium salts are precipitated during this heat treatment. This results in stabilisation of the milk. The usual heat treatment is at 120–122 °C for several minutes, in tubular or plate heat exchangers. This heat treatment also influences the viscosity of the final product, which is an important quality attribute. The milk is then concentrated by vacuum evaporation. The evaporation temperature is in the range 50–60 °C and is usually carried out in multiple-effect, falling film, evaporator systems. Plate and centrifugal evaporators may also be used. Two or three effects are common, but up to seven effects have been used. The density of the milk is monitored until it reaches a value that corresponds to the desired final composition of the evaporated milk. The concentrated milk is then homogenised in two stages in a pressure homogeniser, operating at 12.5–25.0 MPa. It is then cooled to 14°C. The concentrated milk is then tested for stability by heating samples to sterilisation temperature. If necessary, more stabilising salts, such as disodium or trisodium phosphates, are added to improve stability. The concentrated milk is then filled into containers, usually tinplate cans, and sealed by double seaming. The filled cans are then sterilised in a retort at 110–120°C for 15–20 min. If a batch retort is used, it should have a facility for continually agitating the cans during heating, to ensure that any protein precipitate formed is uniformly distributed throughout the cans. The cans are then cooled and stored at not more than 15°C.

As and alternative to in-package heat processing, the concentrated milk may be UHT treated at 140°C for about 3 s and aseptically filled into cans, cartons or 'bag in box' containers.

Evaporated milk has been used as a substitute for breast milk, with the addition of vitamin D, in cooking and as a coffee whitener. A lowfat product may be manufactured using skimmed or semiskimmed milk as the raw material. Skimmed milk concentrates are used in the manufacture of ice cream and yoghurt. Concentrated whey and buttermilk are used in the manufacture of margarine and spreads.

## Sweetened Condensed Milk

This product consists of evaporated milk to which sugar has been added. It normally contains about 8% fat, 20% snf and 45% sugar. Because of the addition of the sugar, the water activity of this sweetened concentrate is low enough to inhibit the growth of spoilage and pathogenic microorganisms. Consequently, it is shelf stable, without the need for sterilisation. In the manufacture of sweetened condensed milk, whole milk is standardised and heat treated in a similar manner to evaporated milk. At this point granulated sugar may be added, or sugar syrup may be added at some stage during evaporation. Usually, evaporation is carried out in a two- or three-stage, multiple-effect, falling film, evaporator at 50–60 °C. Plate and centrifugal evaporators may also be used. If sugar syrup is used, it is usually drawn into the second effect evaporator. The density of the concentrate is monitored during the evaporation. Alternatively, the soluble solids content is monitored, using a refractometer, until the desired composition of the product is attained. The concentrate is cooled to about 30 °C. Finely ground lactose crystals are added, while it is vigorously mixed. After about 60 min of mixing, the concentrate is quickly cooled to about 15 °C. The purpose of this seeding procedure is to ensure that, when the supersaturated solution of lactose crystallises out, the crystals formed will be small and not cause grittiness in the product. The cooled product is then filled into cans, cartons or tubes and sealed in an appropriate manner

Sweetened condensed milk is used in the manufacture of other products such as ice cream and chocolate. If it is to be used for these purposes, it is packaged in larger containers such as 'bag in box' systems, drums or barrels.

Concentrated fruit and vegetable juices are also produced for sale to the consumer.

## Evaporation as a Preparatory Step to Further Processing

One important application for evaporation is to preconcentrate liquids which are to be dehydrated by spray drying, drum drying or freeze drying. The specific steam consumption of such dryers is greater than single-effect evaporators and much greater than multiple-effect systems, particularly, if MVR or TVR is incorporated into one or more of the effects. Whole milk, skimmed milk and whey are examples of liquids which are preconcentrated prior to drying. Equipment similar to that used to produce evaporated milk is used and the solids content of the concentrate is in the range 40–55%.

Instant Coffee: Beverages, such as coffee and tea, are also available in powder form, so called instant drinks. In the production of instant coffee, green coffee beans are cleaned, blended and roasted. During roasting, the colour and flavour develop. Roasting is usually carried out continuously. Different types of roasted beans, light, medium and dark, are produced by varying the roasting time. The roasted beans are then ground in a mill to a particle size to suit the extraction equipment, usually in the range 1000–2000 $\mu m$. The coffee soluble are extracted from the particles, using hot water as the solvent. Countercurrent, static bed or continuous extractors are used. The solution leaving the extractor usually contains 15–28% solids. After extraction, the solution is cooled and filtered. This extract may be directly dried by spray drying or freeze drying. However, it is more usual to concentrate the solution to about 60% solids by vacuum evaporation. Multiple-effect falling film systems are commonly used. The volatile flavour compounds are stripped from the solution, before or during the evaporation, in a similar manner to that used when concentrating fruit juice and added back to the concentrate before drying. Coffee powder is produced using a combination of spray drying and fluidised bed drying. Alternatively, the concentrated extract may be frozen in slabs, the slabs broken up and freeze dried in a batch or continuous freeze drier.

Granulated Sugar: Vacuum evaporation is used in the production of granulated sugar from sugar cane and sugar beet. Sugar juice is expressed from sugar cane in roller mills. In the case of sugar beet, the sugar is extracted from sliced beet, using heated water at 55–85°C, in a multistage, countercurrent, static bed or moving bed extractor. The crude sugar juice, from either source, goes through a series of purification operations. These include screening and carbonation. Lime is added and carbon dioxide gas is bubbled through the juice. Calcium carbonate crystals are formed. As they settle, they carry with them a lot of the insoluble impurities in the juice. The supernatant is taken off and filtered. Carbonation may be applied in two or more stages during the purification of the juice. The juice may be treated with sulphur dioxide to limit nonenzymic browning. This process is known as sulphitation. The treated juice is again filtered. Various type of filters are used in processing sugar juice including plate and frame, shell and tube and rotary drum filters. The purified juice is concentrated up to 50–65% solids by vacuum evaporation. Multiple-effect systems are employed, usually with five effects. Vertical short tube, long tube and plate evaporators are used. The product from the evaporators is concentrated further in vacuum pans or single-effect short tube evaporators, sometimes fitted with an impeller in the downcomer. This is known as sugar boiling. Boiling continues until the solution becomes supersaturated. It is then shocked by the addition of a small amount of seeding material, finely ground sugar crystals, to initiate crystallisation. Alternatively, a slurry of finely ground sugar crystals in isopropyl alcohol may be added at a lower degree of supersaturation. The crystals are allowed to grow, under carefully controlled conditions, until they reach the desired size and number. The slurry of sugar syrup and crystals is discharged from the evaporator into a temperature-controlled tank, fitted with a slow-moving mixing element. From there it is fed to filtering centrifugals, or basket centrifuges where the crystals are separated from the syrup and washed. The crystals are dried in heated air in rotary driers and cooled. The dry crystals are conveyed to silos or packing rooms. The sugar syrup from the centrifuges is subjected to further concentration, seeding and separation processes, known as second and third boilings, to recover more sugar in crystal form.

## The use of Evaporation to Reduce Transport, Storage and Packaging Costs

Concentrated Fruit and Vegetable Juices: Many fruit juices are extracted, concentrated by vacuum evaporation, and the concentrate frozen on one site, near the growing area. The frozen concentrate is then shipped to several other sites where it is diluted, packaged and sold as chilled fruit juice. Orange juice is the main fruit juice processed in this way. The fruit is graded, washed and the juice extracted using specialist equipment described in. The juice contains about 12.0% solids at this stage. The extracted juice is then finished. This involves removing bits of peel, pips, pulp and rag from the juice by screening and/or centrifugation. The juice is then concentrated by vacuum evaporation. Many types of evaporator have been used, including high vacuum, low temperature systems. These had relatively long residence times, up to 1 h in some cases. To inactivate enzymes, the juice was pasteurised in plate heat exchangers before being fed to the evaporator. One type of evaporator, in which evaporation took place at temperatures as low as 20 °C, found use for juice concentration. This operated on a heat pump principle. A refrigerant gas condensed in the heating jacket of the calandria, releasing heat, which caused the liquid to evaporate. The liquid refrigerant evaporated in the jacket of the condenser, taking heat from the water vapour, causing it to condense.

Modern evaporators for fruit juice concentration, work on a high temperature, short time (HTST) principle. They are multiple-effect systems, comprised of up to seven falling film or plate evaporators. The temperatures reached are high enough to inactivate enzymes, but the short residence times limit undesirable changes in the product. They are operated with forward flow or mixed flow feeding patterns. Some designs of this type of evaporator are known as thermally accelerated short time or TASTE evaporators.

Volatile compounds, which contribute to the odour and flavour of fruit juices, are lost during vacuum evaporation, resulting in a concentrate with poor organoleptic qualities. It is common practice to recover these volatiles and add them back to the concentrate. Volatiles may be stripped from the juice, prior to evaporation, by distillation under vacuum. However, the most widely used method is to recover these volatiles after partial evaporation of the water. The vapours from the first effect of a multiple-effect evaporation system consist of water vapour and volatiles. When these vapours enter the heating jacket of the second effect, the water vapour condenses first and the volatiles are taken from the jacket and passed through a distillation column, where the remaining water is separated off and the volatiles are concentrated.

Orange juice is usually concentrated up to 65% solids, filled into drums or 'bag in box' containers and frozen in blast freezers. Unfrozen concentrated juice may also be transported in bulk in refrigerated tankers or ships' holds. Sulphur dioxide may be used as a preservative. Concentrated orange juice may be UHT treated and aseptically filled into 'bag in box' containers or drums. These concentrates may be diluted back to 12% solids, packed into cartons or bottles and sold as chilled orange juice. They may also be used in the production of squashes, other soft and alcoholic drinks, jellies and many other such products.

A frozen concentrated orange juice may also be marketed as a consumer product. This usually has 42–45% solids and is made by adding fresh juice, cutback juice, to the 65% solids concentrate, together with some recovered volatiles. This is filled into cans and frozen. Concentrate containing 65% solids, with or without added sweetener, may be pasteurised and hot-filled into cans, which are rapidly cooled. This is sold as a chilled product. UHT treated concentrate may also be aseptically filled into cans or cartons for sale to the consumer from the ambient shelf.

Juices from other fruits, including other citrus fruits, pineapples, apples, grapes, blackcurrants and cranberries, may be concentrated. The procedures are similar to those used in the production of concentrated orange juice. When clear concentrates are being produced, enzymes are used to precipitate pectins, which are then separated from the juice.

Vacuum evaporation is also applied to the production of concentrated tomato products. Tomatoes are chopped and/or crushed and subjected to heat treatment, which may be the hot break or cold break process. Skin and seeds are removed and the juice extracted in a cyclone separator. The juice is then concentrated by vacuum evaporation. For small-scale production, vacuum pans may be employed. For larger throughputs, two- or three-effect tubular or plate evaporators may be used. If highly concentrated pastes are being produced, agitated thin film evaporators may be used as finishers.

Glucose syrup, skimmed milk and whey are among other food liquids that may be concentrated, to reduce weight and bulk, and so reduce transport and storage costs.

## Distillation

Distillation is a separation process, separating components in a mixture by making use of the fact that some components vaporize more readily than others. When vapours are produced from a mixture, they contain the components of the original mixture, but in proportions which are determined by the relative volatilities of these components. The vapour is richer in some components, those that are more volatile, and so a separation occurs.

In fractional distillation, the vapour is condensed and then re-evaporated when a further separation occurs. It is difficult and sometimes impossible to prepare pure components in this way, but a degree of separation can easily be attained if the volatilities are reasonably different. Where great purity is required, successive distillations may be used. The following types of distillation processes are in use:

## Steam Distillation

In some circumstances in the food industry, distillation would appear to be a good separation method but it cannot be employed directly as the distilling temperatures would lead to breakdown of the materials. In cases in which volatile materials have to be removed from relatively non-volatile materials, steam distillation may sometimes be used to effect the separation at safe temperatures.

A liquid boils when the total vapour pressure of the liquid is equal or more than the external pressure on the system. Therefore, boiling temperatures can be reduced by reducing the pressure on the system; for example by boiling under a vacuum, or by adding an inert vapour which by contributing to the vapour pressure, allows the liquid to boil at a lower temperature. Such an addition must be easily removed from the distillate, if it is unwanted in the product, and it must not react with any of the components that are required as products. The vapour that is added is generally steam and the distillation is then spoken of as steam distillation.

If the vapour pressure of the introduced steam is $p_s$ and the total pressure is $P$, then, the mixture will boil when the vapour pressure of the volatile component reaches a pressure of $(Pp_s)$, compared with the necessary pressure of $P$ if there were no steam present. The distribution of steam and the volatile component being distilled, in the vapour, can be calculated. The ratio of the number of molecules of the steam to those of the volatile component, will be equal to the ratio of their partial pressures–

$$p_A / p_s = (P - p_s) / p_s = (w_A / M_A) / (w_s / M_s)$$

and so the weight ratios can be written:

$$w_A / w_S = (P - p_s) / p_s \times (M_A / M_s)$$

where $p_A$ is the partial pressure of the volatile component, $p_s$ is the partial pressure of the steam, $P$ is the total pressure on the system, $w_A$ is the weight of component $A$ in the vapour, $w_s$ is the weight of steam in the vapour, $M_A$ is the molecular weight of the volatile component and $M_s$ is the molecular weight of steam.

Very often the molecular weight of the volatile component that is being distilled is much greater than that of the steam, so that the vapour may contain quite large proportions of the volatile component. Steam distillation is used in the food industry in the preparation of some volatile oils and in the removal of some taints and flavours, from edible fats and oils.

## Batch Distillation

Batch distillation is the term applied to equipment into which the raw liquid mixture is admitted and then boiled for a time. The vapours are condensed. At the end of the distillation time, the liquid remaining in the still is withdrawn as the residue. In some cases the distillation is continued until the boiling point reaches some predetermined level, thus separating a volatile component from a less volatile residue. In other cases, two or more fractions can be withdrawn at different times and these will be of decreasing volatility. During batch distillation, the concentrations change both in the liquid and in the vapour.

Let $L$ be the mols of material in the still and $x$ be the concentration of the volatile component. Suppose an amount $dL$ is vaporized, containing a fraction $y$ of the volatile component.

Then writing a material balance on component $A$, the volatile component:

$$y \Delta L = \Delta (Lx) = L\Delta x + x\Delta L$$
$$y \Delta L - x\Delta L = L\Delta x$$
$$(y - x)\Delta L = L\Delta x$$
$$\Delta L / L = \Delta x / (y - x)$$

and this is to be integrated from $L_0$ moles of material of concentration $x_0$ up to $L$ moles at concentration $x$.

To evaluate this integral, the relationship between $x$ and $y$, that is the equilibrium conditions, must be known.

If the equilibrium relationship is a straight line, $y = mx + c$, then the integral can be evaluated:

$$\mathrm{Log}_e L / L_0 = 1 \mathrm{Log}_e (m - 1) x + c$$
$$(m - 1) \ (m - 1) x_0 + c$$

or

$$L / L_0 = [(y - x)/(y_0 - x_0)]^{1/(m-1)}$$

In general, the equilibrium relationship is not a straight line, and the integration has to be carried out graphically. A graph is plotted of $x$ against $1/(y - x)$, and the area under the curve between values of $x_0$ and $x$ is measured.

## Vacuum Distillation

Reduction of the total pressure in the distillation column provides another means of distilling at lower temperatures. When the vapour pressure of the volatile substance reaches the system pressure, distillation occurs. With modern efficient vacuum-producing equipment, vacuum distillation is tending to supplant steam distillation. In some instances, the two methods are combined in vacuum steam distillation.

## Distillation Equipment

The conventional distillation equipment for the continuous fractionation of liquids consists of three main items: a boiler in which the necessary heat to vaporize the liquid is supplied, a column in which the actual contact stages for the distillation separation are provided, and a condenser for condensation of the final top product.

The condenser and the boiler are straightforward. The fractionation column is more complicated as it has to provide a series of contact stages for contacting the liquid and the vapour. The conventional arrangement is in the form of "bubble-cap" trays, The vapours rise through the bubble caps. The liquid flows across the trays past the bubble caps where it contacts the vapour and then over a weir and down to the next tray. Each tray represents a contact stage, or approximates to one as full equilibrium is not necessarily attained, and a sufficient number of stages must be provided to reach the desired separation of the components.

In steam distillation, the steam is bubbled through the liquid and the vapours containing the volatile component and the steam are passed to the condenser. Heat may be provided by the condensation of the steam, or independently. In some cases the steam and the condensed volatile component are immiscible, so that separation in the condenser is simple.

## Application

## Manufacture of Whisky

Whisky is a spirit produced by the distillation of a mash of cereals, which may include barley, corn, rye and wheat, and is matured in wooden casks. There are three types of Scotch and Irish whisky: malt whisky produced from 100% malted (germinated) barley, grain whisky produced from unmalted cereal grains and blended whisky which contains 60– 70%grain whisky and 30–40% malt whisky.

## Manufacture of Neutral Spirit

A multicolumn distillation plant is used for producing neutral spirits from fermented mash. A typical system would be comprised of five columns: a whisky-separating column, an aldehyde column, a product-concentrating column, an aldehyde-concentrating column and a fusel oils concentrating column.The whisky-separating column is fitted with sieve plates, with some bubble cap plates near the top of the column. The other four columns are fitted with bubble cap plates. The fermented mash containing 7% (v/v) of alcohol is fed to near the top of the whisky-separating column. The overhead distillate from this column is fed to the aldehyde column. The bottom product from this column is pumped to the middle of the product-concentrating column. The end product, neutral spirit, is withdrawn from near the top of this column.

## Recovery of Solvents from Oil after Extraction

Most of the solvent can be recovered by evaporation using a film evaporator. However, when the solution becomes very concentrated, its temperature rises and the oil may be heat-damaged. The last traces of solvent in the oil may be removed by steam distillation or stripping with nitrogen.

## Concentration of Aroma Compounds from Juices and Extracts

By evaporating 10–30% of the juice in a vacuum evaporator, most of the volatile aroma compounds leave in the vapour. This vapour can be fed to a distillation column. The bottom product from the column is almost pure water and the aroma concentrate leaves from the top of the column. This is condensed and may be added back to the juice or extract prior to drying. Fruit juices and extract of coffee may be treated in this way.

## Extraction of Essential Oils from Leaves and Seeds

This may be achieved by steam distillation. The material in a suitable state of subdivision is placed on a grid or perforated plate above heated water. In some cases the material is in direct contact with the water or superheated steam may be used. If the oil is very heat sensitive distillation may be carried out under vacuum.

# Dehydration

Dehydration in food processing is a means by which many types of food can be preserved for indefinite periods by extracting the moisture, thereby inhibiting the growth of microorganisms. Dehydration is one of the oldest methods of food preservation and was used by prehistoric peoples in sun-drying seeds. The North American Indians preserved meat by sun-drying slices, the Chinese dried eggs, and the Japanese dried fish and rice.

Hot-air dehydration was developed in France in 1795. Modern dehydration techniques have been largely stimulated by the advantages dehydration gives in compactness; on the average, dehydrated food has about 1/15 the bulk of the original or reconstituted product. The need to transport large shipments of food over great distances during World War II provided much of the stimulus to perfect dehydration processes. The advantages of reduced bulk later came to be appreciated by campers and backpackers and also by relief agencies that provide food in times of emergency and disaster.

Dehydration equipment varies in form with different food products and includes tunnel driers, kiln driers, cabinet driers, vacuum driers, and other forms. Compact equipment suitable for home use is also available. A basic aim of design is to shorten the drying time, which helps retain the basic character of the food product. Drying under vacuum is especially beneficial to fruits and vegetables. Freeze-drying benefits heat-sensitive products by dehydrating in the frozen state without intermediate thaw. Freeze-drying of meat yields a product of excellent stability, which on rehydration closely resembles fresh meat.

The dairy industry is one of the largest processors of dehydrated food, producing quantities of whole milk, skim milk, buttermilk, and eggs. Many dairy products are spray dried—that is, atomized into a fine mist that is brought into contact with hot air, causing an almost instant removal of moisture content.

# Smoking

Smoking in food processing, the exposure of cured meat and fish products to smoke for the purposes of preserving them and increasing their palatability by adding flavour and imparting a rich brown colour. The drying action of the smoke tends to preserve the meat, though many of the chemicals present in wood smoke (e.g., formaldehyde and certain alcohols) are natural preservatives as well.

Pastrami sandwich; rye bread.

Pastrami sandwich, traditionally made from beef brisket or navel that has been cured in brine, seasoned with a spice rub, slow-smoked, and then steamed, before being sliced and served hot on rye bread.

Smoking is one of the oldest of food preservation methods, probably having arisen shortly after the development of cooking with fire. The practice attained high levels of sophistication in several cultures, notably the smoking of fish in Scandinavia and northwestern North America and the production of smoked hams in Europe and the United States. Interest in smoking meats, which had declined during the mid-20th century owing to the popularity of chemical preservatives, was revived late in the century by the so-called natural or health food movement.

Whether done on a commercial or a home scale, the smoking technique involves hanging the meat or placing it on racks in a chamber designed to contain the smoke. Commercial smokehouses, usually several stories high, often use steampipes to supplement the heat of a natural sawdust fire. Hickory sawdust is the preferred fuel. Whatever the size of the smoking operation, it is imperative that a hardwood fire be used. The softwood of conifers such as spruce and pine contains pitch, which produces a film on the meat and imparts a bitter taste. Generally, smokehouse temperatures vary from 109 to 160 °F (43 to 71 °C), and smoking periods vary from as short as a few hours to as long as several days, depending on the type of meat and its moisture content. After smoking, the meat is chilled as rapidly as possible and cut and wrapped for the retail trade.

In the United States, pork and beef hams, bacon bellies, and sausages are the most common commercially smoked meats. However, amateurs using ordinary smoke ovens or adapting barbecue grills to the purpose have successfully used the smoking technique to flavour and preserve not only meat, fowl, and fish but also cheeses, nuts and seeds, hard-boiled eggs, and berries, as well as the variety meats including heart, tongue, and liver.

In order to shorten the production process, commercial meats are sometimes artificially "smoked" by dipping them in a solution of preservative chemicals or by painting them with such a solution. But because this procedure involves no natural drying action, it has practically no preservative effect.

## Extrusion Cooking

Food extrusion is defined as "A process in which material is pushed through an orifice or a die of given shape, the pushing force is applied using a piston or a screw, In food applications, screw extrusion is predominant".

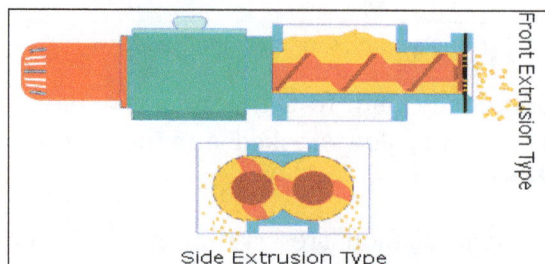

Extruder from front and side view.

Historically, one can trace the use of a screw as a conveying device to the greek phi-losopher Archimedes, who used a single screw in a cylindrical open channel to pump water uphill. Today's extruder consists of one or more screws encased in a metal barrel, attached to a drive motor. A hopper at one end is used to feed raw materials, while a die on the other gives shape to the product. Extruders were developed in the 1870s to manufacture sausage. Application of the single screw extruder evolved during the 1930s, when it was used to mix semolina flour and water to make pasta products. It was also used in the process of making ready-to-eat (RTE) cereals to shape hot, pre-cooked dough. In both of these applications, the level of shear rate was low. During the late 1930s and 1940s, directly expanded corn curls were made using extruders, which were characterized by extremely high shear rates. The first patent on an application of twin screw extrusion technology was filed in mid-1950s. Since then, the application of extrusion technology has widened and grown dramatically.

## Extruders

Extruders come in several designs, dependent upon their application. Some extruders are designed simply to convey the raw materials, while others are designed to mix and knead them; most, however, are designed to impart mechanical and thermal energy to the raw materials to bring about desired physico-chemical changes.

Extruders can be broadly categorized on the basis of:

## Method of operation

- Cold Extruders.

- Hot Extruders (Extrusion Cooking).

## Method of construction

- Single-or twin- screw extruders.

The most commonly used extruders are single and twin-screw. Extruders with more than two screws have been used in the plastics industry but not in food processing.

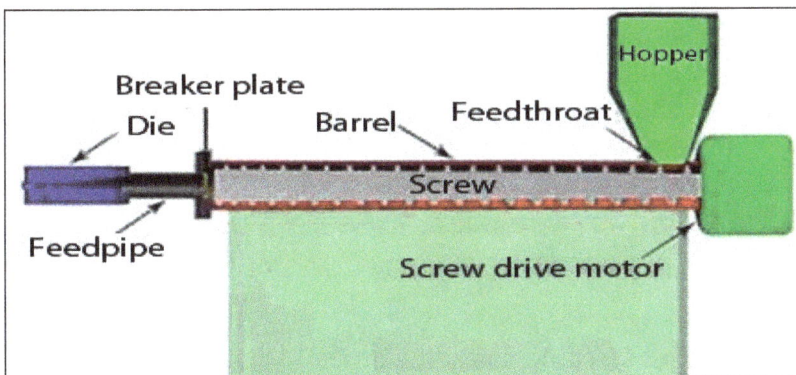

Different parts of extruder.

Extruders are composed of five main parts:

- Pre-conditioning system;
- Feeding system;
- Screw or worm;
- Barrel;
- Die and the cutting mechanism.

Also, they can vary with respect to screw, barrel and die configuration. The selection of each of these items will depend on the raw material used and the final product desired.

## Pre-conditioning

Pre-conditioning with steam or water has always been an important part of the extrusion process. Pre-conditioning is not applied to all extrusion processes. In general, this step is applied when moisture contents around 20 to 30% and long residence times of the material are used. Pre- conditioning favours uniform particle hydration, reduces retention times within the extruder and increases throughput, increasing the life of the equipment, due to a reduction in the wearing of barrel and screw components, also reducing the costs of energy involved in the process.

## Feeding System

It is necessary to guarantee a constant and non-interrupted feeding of the raw materials into the extruder for an efficient and uniform functioning of the extrusion process.

## Screw

The screw of the extruder is certainly its most important component, not only to determine cooking degree, gelatinization and dextrinization of starch and protein denaturation, but also to ensure final product quality. Screws can be mono-piece (composed of a unique piece) or multi- piece (composed of various elements) Screw elements can vary in number and shapes, each segment is designed for a specific purpose. Some elements only convey raw or pre-conditioned material into the extruder barrel, while other segments compress and degas the feedstock. Others must promote kneading, backflow and shear.

## Barrel or Sleeves

The barrel is divided into feeding, kneading anThe sleeves surrounding the screw can be solid, but they are often jacketed to permit circulating of steam or superheated oil for heating or water or air for cooling, thus enabling the precise adjustment of the temperature in the various zones of the extruder.

## Die

The die presents two main functions: give shape to the final product and promote resistance to material flow within the extruder permitting an increase in internal pressure. The die can present various designs and number of orifices.

## Cutting Mechanism

The cutting mechanism must permit obtaining final products with uniform size. Product size is determined by the rotation speed of the cutting blades. This mechanism can be horizontal or vertical.

## Single Screw Extruder

Single screw extruders contain a single rotating screw in a metal barrel, and come in varying patterns. The most commonly used single screws have a constant pitch. The raw materials are fed in a granular form at the hopper located in the feed section. The rotating action of the screw conveys the material to the transition section. In the transition section, the screw channel becomes shallower and the material is compacted. A major portion of mechanical energy is dissipated in this section, which results in a rise in temperature of the material. Starch becomes gelatinized, and the material becomes more cohesive. It is transported further by the metering section and pushed through the die opening. The barrels of single-screw extruders usually have helical or axial grooves on inner surfaces. This helps to convey and mix the material more effectively.

## Twin Screw Extruder

Twin-screw extruders are composed of two axis that rotate inside a single barrel; usually the internal surface of the barrel of twin-screw extruders is smooth. However, this type of extruder is little used in the food industry, even though they present more efficient displacement propertiesWhen the material enters the barrel, the ingredients are thoroughly mixed before further processing in the other zones of the extruder. In this initial step, the screw is designed with a large screw channel depth to provide enough space between the root of the screw and the barrel for sufficient mixing to take place, and often, the screws are reverse-threaded to permit intensive mixing and longer residence times before delivery. In the next zone, the diameter of the root increases rapidly while the channel depth becomes shallower in order to provide material cooking, thus increasing the pressure applied to the product, and the starchy content of food is gelatinized and the proteinaceous material denatured.

Table: Difference between single and twin screw extruder.

| PARAMETERS | SINGLE SCREW EXTRUDER | TWIN SCREW EXTRUDER |
| --- | --- | --- |
| Transport mechanism | Friction between metal and food material | Positive displacement |
| Throughput capacity | Dependent on moisture, fat content and pressure. | Independent |
| Approx. power consumption per kg of product | 900-1500 KJ | 400-600 KJ |
| Heat distribution | Large temperature difference(20-180°C) | Small temperature difference(110-180°C) |
| Mechanical power dissipation | Large shear forces (550-6000kpa) | Small shear forces (2000-4000kpa) |
| De-gassing possibilities | Simple | Difficult |
| Rigidity | High | Bearing capacity is vulnerable |
| Capital cost | low | High (twice) |
| Minimum water content | 10% | 8% |
| Maximum water content | 30% | 95%    Chessari [363] (2001) |

## Physico-chemical Changes during Extrusion

The most used raw materials in the extrusion process are starch and protein based ma
terials. The structure of the extruded products may be formed from starch or protein
polymers. Most products, such as breakfast cereals, snacks and biscuits are formed
from starch, while protein is used to produce products that have meat-like character
istics and that are used either as full or partial replacements for meat in ready meals,
dried foods and many pet food products. Dehgon- shoar *et al.*

Major changes occurs during extrusion process are:
- Changes in starch.
- Changes in proteins.
- Changes in lipid.
- Changes in fibre.

## Raw Materials

In general, the chemical or physicochemical changes in biopolymers that can occur
during extrusion cooking include: binding, cleavage, loss of native conformation, frag-
mention, recombination and thermal degradation. The structure of an extruded product is created by forming a fluid melt from a polymer and blowing bubbles of water
vapour into the fluid to form a foam. The bubbles rapidly expand as the superheated
water is released very quickly at atmospheric pressure.

## Changes in Starches

The major difference between extrusion processing and conventional food processing is that in the former starch gelatinization occurs at much lower moisture contents (12-22%). Once inside the extruder, and at relatively high temperatures, the starch granules melt and become soft, besides changing their structure that is compressed to a flattened form. The application of heat, the action of shear on the starch granule and water content destroy the organized molecular structure, also resulting in molecular hydrolysis of the material. The starch polymers are then dispersed and degraded to form a continuous fluid melt. The fluid polymer continuum retains water vapour bubbles and stretches during extrudate expansion until the rupture of cell structure. The starch polymer cell walls recoil and stiffen as they cool to stabilize the extrudate structure. Finally, the starch polymer becomes glassy as moisture is removed, forming a hard brittle texture.

## Changes in Protein

Proteins are biopolymers with a great number of chemical groups when compared to polysaccharides and are therefore more reactive and undergo many changes during the extrusion process, with the most important being denaturation. During extrusion, disulfide bonds are broken and may re-form. Electrostatic and hydrophobic interactions favour the formation of insoluble aggregates. The creation of new peptide bonds during extrusion is controversial. High molecular weight proteins can dissociate into smaller subunits. Enzymes, also proteins, lose their activity after being submitted to the extrusion process due to high temperatures and shear.

## Changes in Lipids

Fats and oils can be described as lipids. Lipids have a powerful influence in extrusion cooking processes by acting as lubricants, because they reduce the friction between particles in the mix and between the screw and barrel surfaces and the fluid melt.

## Changes in Fibres

Research has shown that cooking fibres by extrusion can produce changes in their structural characteristics and physicochemical properties, with the main effect being a redistribution of insoluble fibre to soluble fibre. This effect would be the result of the rupture of covalent and non- covalent bonds between carbohydrates and proteins associated to the fibre, resulting in smaller molecular fragments, that would be more soluble.

## Influence of Extrusion Cooking on Moisture and Temperature

In the extrusion process of expanded products with low moisture, the expansion of the final product is inversely related to the moisture of the raw material and directly related to theincrease in extrusion temperature; however, the effect of moisture is more significant. In high moisture extrudates, expansion occurs when the product exits the die, but the structure collapses before the necessary cooling, resulting in a dense and hard product. Another important parameter for extrudate expansion is process temperature. Products do not expand if the temperature does not reach 100°C. Expansion increases with the increase in temperature when moisture content of the material is close to 20%, due to lower viscosity, permitting a more rapid expansion of the molten mass, or due to an increase in water vapour pressure. At low extrusion temperatures, expansion is reduced because starch is not completely molten. Radial expansion degree is proportional to temperature up to a certain value, decreasing at much higher temperatures. The reduction of expansion at very high temperatures is attributed to an increase in dextrinization, weakening starch structure.

## Influence of Extrusion Cooking on Various Parameters

Three main factors of extruded product on consumer view are:

- Product quality.
- Nutritional quality.
- Microbiological quality.

## Influence on Product Quality

Extrusion-cooking has an important influence on product quality, emphasizing features like expansion, texture, shelf-life, colour and flavour. Products obtained with high temperatures and short extrusion process times normally present a porous, open structure, what confers to them a "crunchy" texture. Colour in extruded products is influenced by temperature, raw material composition, residence time, pressure and shear force.

## Effect of Extrusion on Nutrition Quality

The effects of extrusion cooking on nutritional quality are ambiguous. Benefits include destruction of anti-nutritional factors, gelatinization of starch, increased soluble dietary fibre and reduction of lipid oxidation. Starch digestibility is largely dependent on complete gelatinization. High starch digestibility is essential for specialized nutritional foods such as infant and weaning foods. Creation of resistant starch by extrusion may have value in reduced calorie products. The nutritional value of vegetable proteins is generally enhanced by mild extrusion cooking conditions due to the increase in digestibility, probably a result of protein denaturation and the inactivation of enzyme inhibitors present in raw materials, by the exposure of new active sites for enzyme attack.

## Influence on Microbiological Quality

One of the most important consumer requirements is the microbiological safety of food products. Most conventional extruded products such as snack foods and breakfast cereals are safe to eat because the raw materials are subjected to high temperatures (higher than 130°C) and the water activity temperatures (0.1-0.4) of the product is low because the product is dried to a moisture content of less than five per cent. Although it is well known that most vegetative organisms, yeast and moulds are destroyed under typical extrusion conditions (55-145°C).

## Recent Developments in Dairy, Breakfast Cereal and Pet Food Industry

Extruders permit the production of many foods of nutritional importance. The ability of extruders to blend diverse ingredients to make novel foods.

Some of the recent advances under:

1. Dairy

2. Breakfast Cereal and

3. Pet Foods are discussed further

## Extrusion in Dairy Industry

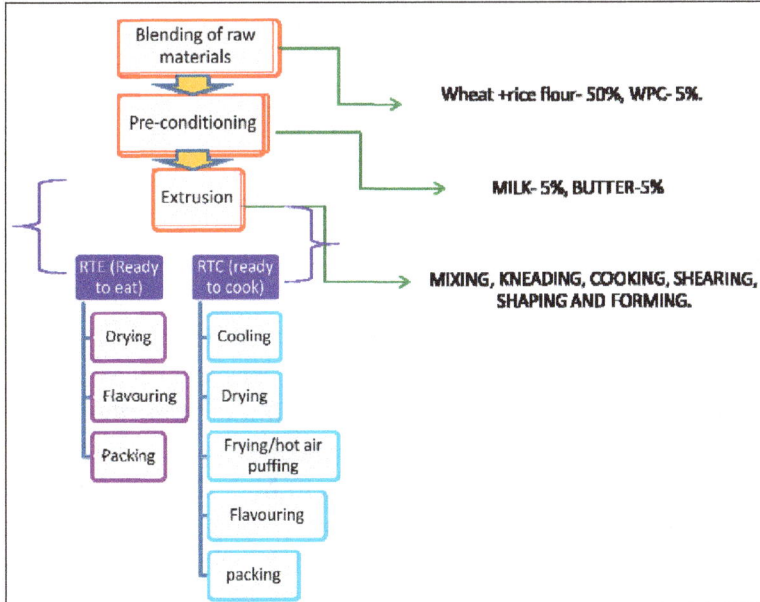

Process of RTE and RTC extruded products.

Milk protein posses health benefits and desirable functional properties. When protein is subjected for mechanical shear, considerable changes in the molecular structures of the protein is seen. This changes leads to a formation of new protein-based food product known as "TEXURIZATION". Texturization stretches and shears the protein to form a new fibre-bundle like structure which withstands - hydration, cooking and other procedures.

## Extrusion for Breakfast Cereals

Basic processes of making breakfast cereals include flaking, oven, gun puffing, baking, shredding and direct expansion (extrusion cooking). Which convert raw and dense grains (7.7 kg/100 cm³) into friable, crunchy or chewable products with the density of 0.6 to 1.6 kg/100 cm³. But extrusion process combines all the above processes and presents advantages over the conventional processes.

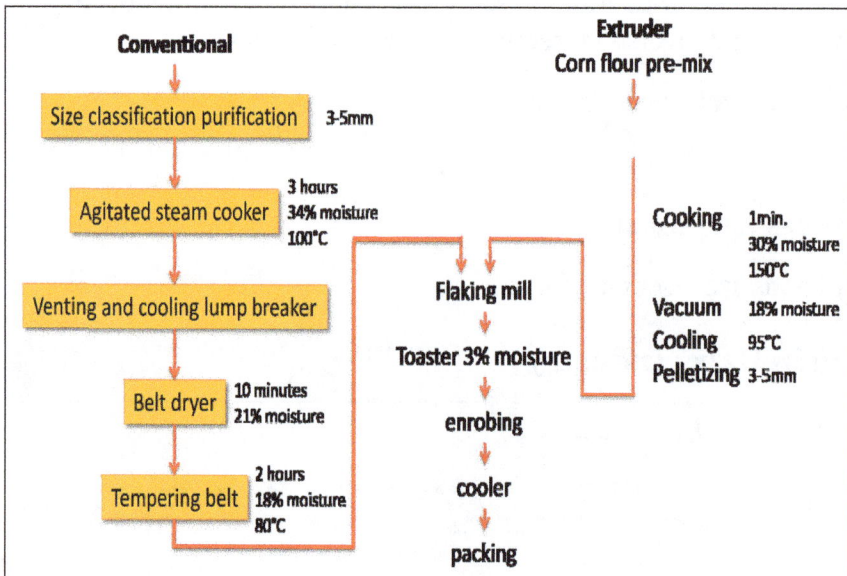

Differences between conventional and extrusion process.

## Meat Analogues

Increase in vegetarians, lead to a preparation of protein incorporated meat extenders and meat analogues which can be obtained from extrusion process. Vegetable protein posses a charecteristic appearance and texture similar to fibrilar structure of meat. Thus, Meat extenders are obtained from thermoplastic extrusion at low moisture contents (20-35%) and Meat analogues are obtained at high moisture contents (50- 70%). The main vegetable proteins used to prepare meat analogues are legumes, soybeans, common beans and peas. The cereals which contain wheat proteins responsible for gluten network.

## Pet foods and Animal Feeds

Extrusion is not only used to prepare food for humans but also used to prepare semi-moist and dry expanded pet foods, aquatic food, and foods for laboratory animals. Whereas, dog and cat foods are directly extruded and dried. Feed for ornamental fish, high-grade complete feeds to maintaining the health, foods for exotic species in aquariums can also be made from extrusion process. This permits better utilization of available cereal grains, vegetable and animal proteins.

# Frying

Frying is an old and widely used method of cooking and processing food. Typically, a food is immersed in heated oil for a short duration in a process known as immersion-oil frying. Numerous types of edible oils of plant and animal origin are used in frying, depending on regional availability. Palm oil is often used in Southeast Asia, coconut and groundnut oil in the Indian subcontinent, and olive oil in the Mediterranean region. During the last five decades, the Western food industry has become increasingly dependent on the frying process to manufacture a variety of snack foods. Fried foods such as potato chips, french fries, and fried fish and chicken have gained worldwide popularity. In this article, the term frying is used to describe a process in which a food is cooked by immersion in heated oil. Alternatively, this process is also referred to as immersion-oil frying and deep fat frying.

A wide variety of products are fried on an industrial scale. Among the most popular are potato chips (also called potato crisps), expanded snacks, roasted nuts, french fries (also called potato chips), extruded noodles, doughnuts, and frozen foods covered with batter, such as fish and chicken.

## Physico-chemical Changes in Foods during Frying

The immense popularity of fried foods is due to the unique flavor and texture imparted to food as it is fried. Many fried foods typically have a porous, crispy outer crust layer with unique flavor but are also soft and moist internally. With starchy foods, it is generally agreed that the crust layer develops as starches in the food gelatinize and the outer layer rapidly dries. However, the kinetics and mechanisms involved in crust formation and related structural changes have yet to be well understood. These mechanisms are dependent on the physical and chemical properties of the food material and the oil used in frying. The complexity of the frying process can be seen in many aspects. For instance, the structure and thickness of the porous crust layer is dependent on the product composition, the processing time, temperature, and composition of the frying oil and material being fried. Heat transfer between the oil and food surface is largely due to the convection mode, whereas heat flux across the crust layer is characterized by the conduction mode of heat transfer. Inter- and intra-sample irregularities pose a significant problem when trying to understand the basic driving forces involved in the movement of oil in a food during frying. This heterogeneity has led to the development of product-specific, empirical models that yield little in the way of fundamental knowledge. As a result, only limited progress has been made in developing a mechanistic understanding of the frying process.

Frying is often the preferred method of cooking for several reasons. The relatively high temperature of the oil used as a heating medium in the process sharply reduces the cooking time, resulting in desirable sensory characteristics like crispy texture and pleasant aroma. In fried foods, such as potato chips and noodles, a product is fried until most of the moisture is removed and a porous structure is created throughout the product. With French fried potatoes, among other foods, the potato strips are fried for a sufficient duration until the outer region becomes porous and stiff, giving it a crispy texture, while the inner region is cooked and moisture is retained. Chemical changes create desirable aromas. The amount of oil absorbed during frying increases the total fat content and caloric value of a food, of concern when a low fat diet is desired.

## Edible Oils used in Frying Foods

In most cases, these foods have little or no native fat prior to frying. Most of the oil absorbed by a food takes place either during immersion in oil or immediately upon the removal from oil. With nuts and other foods having high oil content, the exchange of oil between the food and that used during frying alters the sensory and nutritional characteristics of the fried product.

Table: Typical oil content of fried foods.

| Product | Oil Content (%) |
| --- | --- |
| Potato chips (also known as crisps) | 33-38 |
| Precooked potato French fries | 10-15 |
| Doughnuts | 20-25 |

In addition to the oils mentioned, many modified, fractionated, or hydrogenated oils are used in frying that provide required performance in terms of process, cost, and quality characteristics.

Table: Different types of oils used in food frying.

| Coconut oil |
| --- |
| Corn oil |
| Palm oil |
| Palm oleine |
| Rapeseed oil (low euricic acid) |
| Soybean oil |
| Sunflower seed oil |

## Physiochemical Changes in Oil during Frying

Oil is an excellent heating medium, because it allows high rates of heat transfer into foods being cooked. However, the frying process also causes a number of chemical and physical changes in the oil. These changes not only influence the heating characteristics, but also bring about changes in the sensory and nutritional characteristics of foods. Physicochemical changes in the oil are due to three factors:

- Oxidative changes due to atmospheric oxygen entering the oil from the exposed surface of the oil to the surrounding atmosphere.

- Hydrolytic changes due to water vapors from the product undergoing frying.

- Thermal changes due to oil being maintained at high temperatures.

Hydrolytic reactions cause cleavage of the ester bond of a lipid, resulting in the formation of fatty acids, glycerol, and mono- and diglycerides. Because oxidative reactions and thermal changes alter the unsaturated constituents of triglycerides, at least one of the acyl radicals of the triglyceride molecule is altered. Table shows some of these changes in oils during frying. The three causative agents—namely, moisture, oxygen, and temperature—have a synergistic effect on the compounds produced during frying. Volatile compounds produced during frying influence the organoleptic quality of the food. Non-volatile compounds in the oil are important, because they migrate into the food undergoing frying and are subsequently ingested. These non-volatile compounds are also the basis of several analytical procedures used to measure alterations in the oil due to frying.

Table: Changes in oil during the frying process.

| Changes | Causing Agent | Resulting compounds |
|---------|---------------|---------------------|
| Thermal alteration | Temperature | • Cyclic monomers <br> • Dimers and polymers |
| Oxidative alteration | Air | • Oxidized monomers, dimers and polymers <br> • Nonpolar dimers and polymers <br> • Volatile compounds (hydrocarbons, aldehydes, ketones, alcohols, acids, etc.) |
| Hydrolytic alteration | Moisture | • Fatty acids <br> • Monoglycerides <br> • Diglycerides <br> • • Glycerol |

The preceding mechanisms occur simultaneously. They depend on different types of foods and on interaction with food components. The reactions taking place in the oil are highly complex and are responsible for variation in the results of analytical methods.

Several changes are observed in oils when used in repeated frying. These changes include:

- Development of unique sensory characteristics of oil and food being fried.

- Change in density and viscosity.

- Change in color, such as darkening due to formation of polar compounds.

- Formation of foam due to polymer formation.

- Specific extinction of 232 and 270 nm of conjugated double bonds.

- Change in fatty acid composition, with increase in saturated acids.

- Increase in acid value due mostly to hydrolytic reactions.

- Decrease in iodine value due to elimination of double bonds resulting from polymerization and other reactions.

Numerous factors are used to assess the quality of edible oils used in frying: oil color, odor and taste, percentage of free fatty acids, peroxide value, iodine, and smoke point. These factors also play an important role in determining the final quality of fried foods.

## Dielectric

The definition of dielectric heating can be stated as – 'the process of heating up material by causing dielectric motion in its molecules using alternating electric fields". All materials are made up of molecules that are composed of atoms.

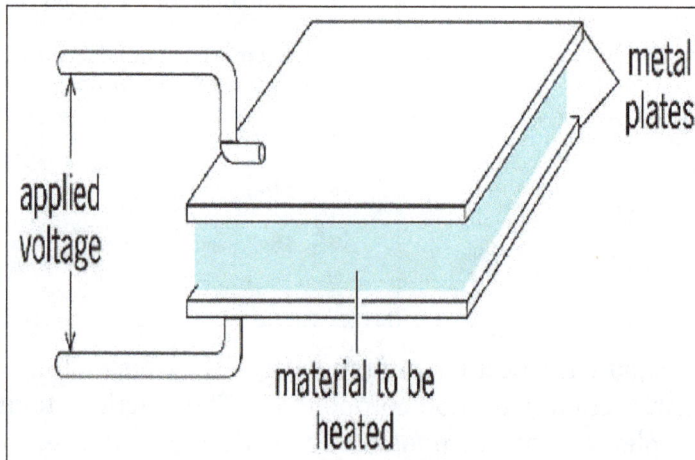

Dielectric Heating.

Polar molecules contain electric dipole moments. When such molecules are exposed to the electric field, they try to align themselves in the direction of the field. When the applied field oscillates, these molecules of the material undergo rotations in order to keep themselves aligned with the field. When the field changes direction, these molecules also reverse their direction. This process is called "Dielectric Rotation".

The temperature of the molecules is related to the kinetic energy of the molecules. In the dielectric rotation of the molecules, as the kinetic energy of the molecules increases, the temperature of the molecules increases. When the molecules collide or come in contact with other molecules, this energy gets transferred to all parts of the material thus heating up the material.

Thus dielectric rotation in the material is often referred to as Dielectric heating of the material. This heating is done using either electric fields of RF frequencies or electromagnetic fields. The applied field should be oscillating for dielectric rotation to take place. The frequency and wavelength of the applied field also affect the functioning of the system.

## Dielectric Heating Working

As described below, the circuit diagram of the dielectric heating system consists of two metals plates to which the electric field is applied. The material to be heated is placed in between these two metals. There are two types of ways in which material are heating using the heating process.

Heating using low-frequency waves, as a near – field effect and heating with high-frequency waves using electromagnetic waves. The type of materials heated using these different types of waves is also different.

Low-frequency waves have higher wavelengths. Thus they can penetrate through non-conductive materials more deeply than electromagnetic waves. The systems using low-frequency fields should have the distance between the radiator and absorber to be less than $1/2\pi$ of the wavelength. So, the process of heating using a low-frequency electric field is near – contact process.

Higher frequency systems have lower wavelengths. Electromagnetic waves and microwaves are used for these systems. In these systems, the distance between metal plates is larger than the wavelength of the applied field. In these systems, conventional far-field electromagnetic waves are formed between the metal plates.

## Applications of Dielectric Heating

Dielectric heating principle using high-frequency electric fields was proposed in the 1930s at Bell Telephone Laboratories. By varying the frequency of electric fields the Dielectric systems are designed for many types of applications.

## When Microwaves are Used

In this dielectric heating, the 2.45GHz of the microwave of frequency is used. Microwave ovens used in homes are an example of this type of applications. These systems provide less penetrative and highly efficient heating system. Microwave volumetric heating provides a larger penetration depth. Thus this heating is used for heating liquids, suspensions, and solids on an industrial scale.

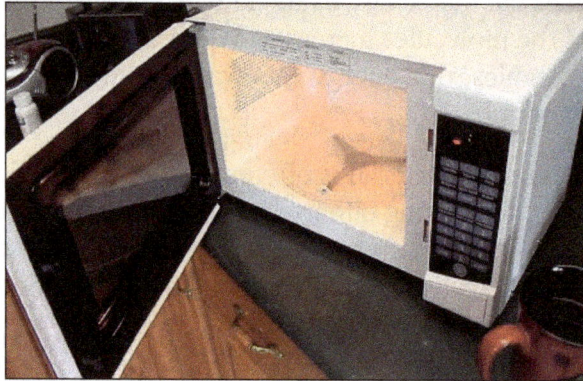

Microwave Oven.

Microwave volumetric heating is applied for pasteurization, Flash pasteurization, microwave chemistry, sterilization, food preservation, biofuel production, etc.

## When Radio Frequencies are Used

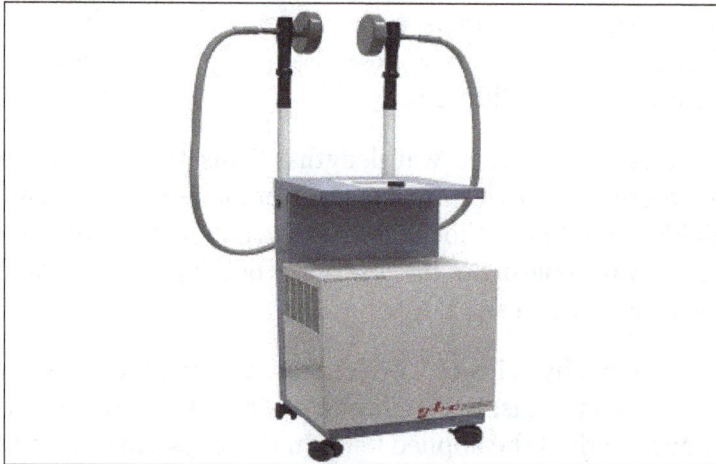

Shortwave Diathermy.

- RF dielectric often finds applications in the crop production area.

- This type of heating is used to kill some pests in food after harvest of the crop.

- This type of heating can heat materials uniformly.

- This type of heating can process food quickly.

- Diathermy, the process of RF heating of muscle for muscle therapy uses this type of heating.

- The process called Hyperthermia therapy, in which higher temperatures are used to kill cancer and tumor tissues, heating with RF frequencies is applied.

## Food Processing

In post-baking of biscuits in the production line, RF dielectric heating will reduce the baking time. Right size, shape and color biscuits can be produced with oven but RF heating can remove the remaining moisture from already dried parts of the biscuits.

- RF heating can increase the capacity of the oven, used in food production factories, up to 50%.

- Cereal-based baby products and breakfast cereals use the post-baking by RF dielectric heating.

- In the drying of food, dielectric baking is used along with conventional baking.

- When an electromagnetic dielectric is used for the baking better quality of food is achieved.

- Nutritional and sensory properties of food can be preserved during food processing when electromagnetic dielectric heating is used, as higher processing temperatures can be achieved in a shorter amount of time.

Right from the period of its invention, dielectric heating is being used in various forms. From an amazing food processor to a precise Electro surgery method, dielectric had found its application in almost every field of science.

Dielectric heating mechanism setup can be viewed as similar to the structure of the capacitor. In capacitor dielectric is placed in between two conducting plates and electricity is produced in a dielectric. Whereas in a dielectric heating system, the material to be heated is placed in between two conducting plates, to which electric field is applied and heat is generated inside the material.

Now a day's dielectric heating has found many applications in the agriculture industry, for implementation of many pest control methods.

## Ohmic

Ohmic heating is often placed among the so-called 'novel food processing technologies'. This is correct if we consider that only in the last 10 to 15 years have the scientific and technical developments achieved finally allowed its safe application at the industrial scale. However, the bases of the technology were set in the late 19th century and the 'Electropure Process' was developed in the 1930's. By 1938 it was used in approximately 50 milk pasteurisers in five US states and served approximately 50,000 consumers.

The amount of energy generated as heat per second (the heating rate) depends on the voltage gradient and on the electrical conductivity of the product ($\sigma$). The most critical property affecting ohmic heating rate is $\sigma$ which is in turn affected by a large number of parameters such as temperature, ionic strength, material microstructure, presence of a second phase and its concentration, presence and concentration of non-electrolytes, etc.

Ohmic heating has been shown, for example, to influence mass transfer properties. The electrical conductivity affects the diffusion from a food sample and at a steady-state temperature, differences in sample $\sigma$ between conventional and ohmic conditions account for the difference in the extent of diffusion.

Several other studies have demonstrated the effect of frequency and wave shape of alternating current on the ohmic heating of foods.

It is possible to have the same heating rate in solid and liquid phases, so overheating of particulates can easily be avoided.

Modern process systems engineering offers a number of tools (namely computer-aided simulation, optimisation and control), which can be used to design new processes or to improve existing ones. These process engineering tools are based on mathematical models and offer a powerful, rational and systematic way of designing and operating food processes. A good example of the benefits of process optimisation can be found in thermal processing (sterilisation, pasteurisation), where it has been shown that new operating policies, computed by suitable optimal control methods, present significant advantages with respect to classical constant-temperature operation. Thus, it can be expected that a similar approach for the case of ohmic heating might also bring a number of improvements and in fact, a number of researchers have investigated the mathematical modelling of ohmic heating, which is certainly a non-trivial task. Different fundamental methodologies for the ohmic process modelling have been followed and reviewed.

The main innovation brought by ohmic heating is the way through which it allows the heating of foods (internal heat generation). This makes all the difference from other commercially available heat processing technologies where heat is either transmitted by conduction or convection (as happens with most types of heat exchanging equipment). Also when compared to microwave heating, for example, ohmic heating presents advantages. In the former, heating is achieved only in a certain depth of the product while the latter heats the entire volume, no matter its size.

Its ability to heat materials rapidly and uniformly is its principal advantage: the over processing of the foods is therefore prevented and so is the further destruction of nutrients and flavour compounds, leading to a higher quality product, both from the nutritional and the organoleptic points of view. Other advantages include:

- The involvement of no hot surfaces: therefore leading to fewer chances of fouling;

- Better controllability: once the rate of heating is governed only by the power supply available (in turn, dependent on the product's electrical conductivity and, in continuous processes, on the flow rate used);

- Less damages caused to shear-sensitive food components: once shear rates are comparable to those existing in conventional pipe work with little additional disturbance to the flow; this leads to low pressure drops in the system;

- High efficiency: once virtually all of the electrical power supplied to the system is dissipated as heat in the product (typical efficiencies of ohmic heaters are above 95 per cent);

- Possibility of handling slurries with high solid contents (products with up to 80 per cent particulates can be processed).

Despite these advantages, there are potential problems with ohmic heating; the main being products with low electrical conductivity values (e.g. those with high fat contents). In this case, ohmic heating might just not be applicable due to the high power input required to process such products. Other limitations include:

- The energy costs of electrical heating systems (these will be depend on the region where it is implemented);

- The limitation imposed by the minimum and maximum values of the electrical conductivity allowed for a given throughput, which in turn will limit the type of products that can be processed by ohmic heating;

- The limitation imposed by the maximum power available from the local electrical supply;

- The requirement for non-conductive materials to be used in linings/housing in order to isolate electrical components.

Ohmic heating can be considered a High Temperature Short Time (HTST) aseptic process, therefore the microbiological safety of the products is not affected (and may even be improved). In fact, the principal mechanisms of microbial inactivation are based on thermal inactivation. It is generally believed that destruction of microorganisms by ohmic heating is due to the thermal effect and electricity plays no role on inactivation. However, the thermal requirement for inactivation is reduced when sub-lethal electrical treatment is applied[21] and recent results clearly show that there is a non-thermal (electrical) effect of the application of ohmic heating on microbial growth.

Similar conclusions are being collected with respect to the non-thermal effects of ohmic heating on food enzymes or vitamins for example, just to quote a few works.

The potential applications of this technique in the food industry are very wide and include blanching, evaporation, dehydration, fermentation, pasteurisation and sterilisation. The advantages discussed above make ohmic heating especially useful for very sticky (viscous) materials or fluids containing solid particles; it may also be used wherever non-uniform heating must be avoided or where mechanical agitation to improve heat transfer is not recommended. A few examples follow.

The thermal processing of fruit purees and jams is traditionally difficult, essentially due to their rheological properties. The problem is aggravated when fruit particles are present in the slurry, as in the case of fruit purees to be incorporated in yoghurts, for example. In fact, in order to process them fully to meet food safety requirements, an over-processing of the liquid phase is necessary. This is mainly due to the heat exchange mechanism used (conduction) and leads to important losses both in nutritional and organoleptic terms. The use of conventional heat exchangers is not possible and scraped surface devices are normally used instead. The contact of the slurry with a hot surface is promoted and mixing is achieved by means of rotating blades. These are responsible for mechanical damage to the fruit particles affecting the final quality of the product and diminishing its acceptability to the consumer. The maintenance of such scraped surface heat exchangers is also more expensive than that of the most usual plate-and-frame or shell-and-tube options. This is one of the cases where ohmic heating can be used with clear advantages over the conventional processes.

The same thermal over-processing problems leading to nutritional and flavour losses are present in the production process of canned soups for instance, which normally integrate food solid particles (e.g. peas, carrot cubes), or even during pet food production, where reported nutritional and flavour losses due to thermal over-processing of the product are considerable.

Despite all the efforts made so far, the application of ohmic heating to food processing is still not fully characterised and not all of its potentialities have been exploited due to the complexity of both the food materials and the phenomena occurring during ohmic heating processing.

## References

- Meaning-thermal-processing- 6790531: sciencing.com, Retrieved 23 January, 2019

- Heat-exchangers-for-the-food-industry: italianfoodtech.com, Retrieved 26 July, 2019

- The-Safety-and-Quality-of-Sous-Vide-Food-327454357: researchgate.net, Retrieved 20 August, 2019

- Pasteurisation-guidelines-food-safety-regulations: foodsafetyhelpline.com, Retrieved 15 June, 2019
- Sterilization, food-preservation: britannica.com, Retrieved 15 May, 2019
- Methods-of-sterilization: apps.who.int, Retrieved 15 January, 2019
- Dehydration-food-preservation: britannica.com, Retrieved 05 February, 2019
- Dielectric-heating-system-working: elprocus.com, Retrieved 02 June, 2019
- Ohmic-heating-in-the-food-industry-610: newfoodmagazine.com, Retrieved 26 March, 2019

# Processing by Removal of Heat

The preservation of food by lowering the temperature helps in maintaining the nutritional value and sensory characteristics of food. Some of the processes used are refrigeration, cooling, freezing, freeze-drying, etc. This chapter closely examines these processes to provide an extensive understanding of the subject.

## Refrigeration

Refrigeration is a science that deals with absorption of heat at a temperature from surroundings at lower temperature and rejecting it to a relatively higher temperature at the cost of some external work following Clausius statement of second law of thermodynamics. Refrigeration sector is one of the largest economic sectors, covering precooling, cooling, air conditioning, and freezing applications for various commodities and people and appears to be an energy intensive sector by consuming about 20 per cent of the total electricity generated worldwide. The refrigeration sector includes air conditioners, household fridges, coolers, pre-coolers, refrigerators, heat pumps, and freezers for various applications, ranging from food cooling to space cooling. Thus, refrigeration has found its importance in many engineering and industrial sectors. One of them is processing, preservation and nourishment of food by storing them in places whose temperature is lower than that of the surroundings. Figure shows the energy consumption of different preservation processes over storage time together with the energy used for the agricultural production.

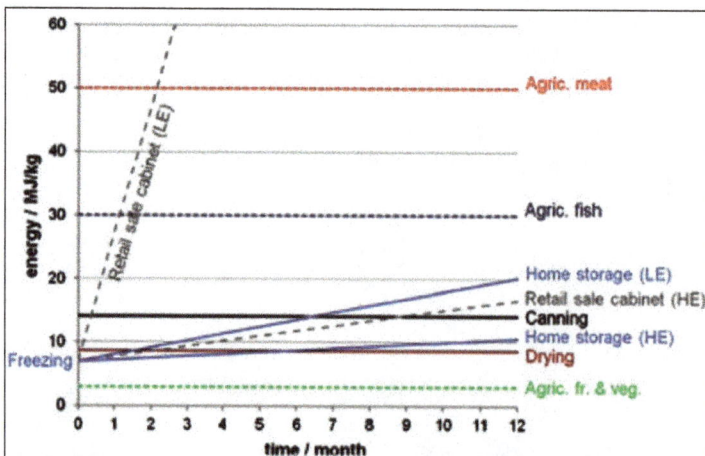

Energy consumption of different preservation processes over storage time together with the energy used for the agricultural production.

## Storage Life of Frozen Foods

According to the International Institute of Refrigeration, storage life is defined as the length of duration of a food product stored under specified conditions so that it is liable for consumption. Different food products have different storage lives. Also, it has to be considered that storage life of a particular food product may vary according to its location and other ecological aspects. The ranges of indicative practical storage lives of various food products are given in table.

Table: Storage life of various food products.

| Product | Optimum Temperature (°C) | Relative Humidity (%) | Storage Life |
|---|---|---|---|
| Free products | | | |
| Apple | -1 to 4.5 | 90-95 | 1-12 months |
| Asparagus | 2 | 95-100 | 2-3 weeks |
| Banana | 13-15 | 90-95 | 1-4 weeks |
| Blueberries | 0.5-1 | 90-95 | 2-3 weeks |
| Broccoli | 0 | 95-100 | 2-3 weeks |
| Cabbage | 0 | 95 | 1 week |
| Cucumber | 7-10 | 95 | 25 weeks |
| Eggplant | 8-12 | 90-95 | 2-3 weeks |
| Ginger | 13 | 65 | 5-10 weeks |
| Guava | 5-10 | 90 | 1-2 weeks |
| Green beans and field peas | 3-7 | 95 | 2-3 months |
| Leafy vegetable | 0 | 95 | 1-3 weeks |
| Lime | 9-10 | 85-90 | 3-5 weeks |
| Mango | 13 | 90-95 | 2-4 weeks |
| Onion | 0 | 70 | 2-3 weeks |
| Papaya | 7-13 | 85-90 | 1-3 weeks |
| Passion fruit | 7-10 | 85-90 | 3-5 weeks |
| Peach | 0 | 95-98 | 2-4 weeks |
| Peppers | 7-10 | 90-95 | 3-5 weeks |
| Potato | 3-4.5 | 90-95 | 5-8 months |
| Sweet corn | 0 | 90-95 | 5-7 days |
| Sweet potato | 13 | 90 | 6-12 months |
| Tomato, pink | 9-10 | 85-95 | 7-14 days |
| Turnip | 0 | 95 | 4-5 months |
| Watermelon | 10-15.5 | 90 | 2-3 weeks |

| Fish | 0-3 | 90 | 1-3 days |
|---|---|---|---|
| Milk | 4 | - | 3-5 days |
| **Cooked Food** | | | |
| Vegetable | 0-4 | - | 2-4 days |
| Fish | 0.4 | - | 2-3 days |
| Meat | 0-4 | - | 3-5 days |
| Soup | 0-4 | - | 2-3 days |

## Different Refrigeration Processes Involved in Food Processing

The principle behind preservation of foods by refrigeration is to reduce and maintain the temperature of the food in order to mitigate any detrimental or undesirable changes occurring in the food. These changes can be microbiological, physiological, biochemical or physical. This in turn can help in improving the nutritional quality of the food. There are several refrigeration processes involved in the processing and nourishment of food. The refrigeration techniques that are implemented one after the other have been illustrated below:

• Pre-cooling: Rapid cooling is necessary to retard the metabolism of food products and to increase their life span. The lesser the temperature is, the better the quality and longevity of the cuisines. Based on the type of the food product, different refrigeration techniques have been proposed. i.e. hydrocooling for small fruits, vacuum cooling for green leafy vegetables, blast-air system for cooling the surfaces of meat products, cooling of milk and other dairy products in specialised tanks, pulsed air in some other products and so on.

• Chilling: After precooling, the food product has to be chilled and maintained at the favourable temperature. Chilling is usually done in classic cold rooms equipped with sufficient ventilation. The chilling and conservation temperatures depend on the sensitivity of the product i.e. for high-sensitivity products such as mangoes, ginger, sweet potatoes temperatures below 8-12 °C are not recommended, as they can endure metabolic disturbances that shorten the life of the goods. For medium-sensitivity products such as tangerines, green beans, potatoes, it is possible to lower the temperature not less than 4-6 °C. For low-sensitivity products, a temperature of 2-3 °C, down to just above the freezing point is recommended. In slaughterhouses, low temperatures that can be achieved by using blast chillers or cold storage rooms are recommended in order to inhibit the growth of microorganisms. For eggs, conservation at higher temperature could be enough as long as the shell remains unbroken and for dairy products an adequate temperature of nearly 8-10 °C would be enough to inhibit the growth of pathogenic germs.

• Freezing: Freezing is a process of lowering the temperature of a product below

its solidification point. Freezing hinders the metabolism of the fruit and vegetable products. Longer the product remaining frozen, longer is its durability for
storing.

- Super-cooling and Super-chilling: These are new processes implemented for
the storage of food products. Supercooling is lowering down the temperature
of a food product just below its freezing point without any formation of ice and
superchilling is partial freezing of a food at a temperature just below its freezing
point.

Apart from the above-mentioned techniques, some other techniques are also adopted
such as crystallisation of fat, cryoseparation of undesirable components; cryoconcen-
tration and so on.

## Refrigerants used in Food Processing Industries

Refrigerants are the working fluids used in refrigeration systems. Selection of refrigerant plays a vital role in any refrigeration industry, so it has been a real challenge to
develop and implement refrigerants having desirable thermodynamic properties and
high energetic and exergetic performances. Various refrigerants used in food processing sectors and their percentage shares have been presented in figure.

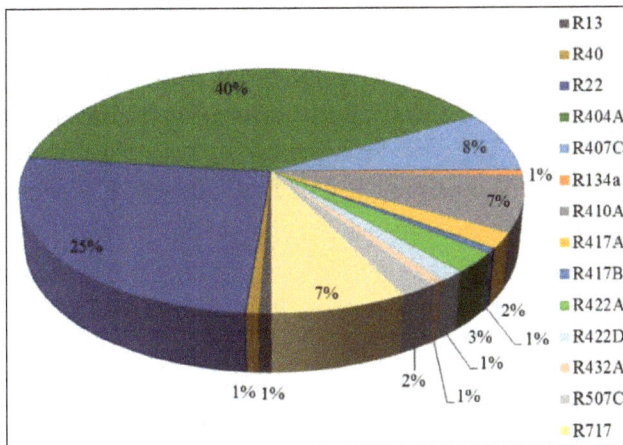

Percentage distribution of refrigerants used in food processing.

Earlier, the refrigerants were selected based on their thermodynamic properties only, but
eventually focus has also been made over the ozone depletion potential (ODP) and global
warming potential (GWP) of refrigerants. Some of the desirable properties of refrigerants needed in food processing industry have been listed below:

- It should have low flammability and toxicity.

- It should zero ODP value and GWP value as minimum as possible.

- It should be easy leak detectable.

Selection of refrigerants has significant impact on safety, reliability, energy consumption, system performance and also on the environment. Some of the refrigerants involved in food processing industries have been listed below in table. However, some alternative refrigerants have now been replaced in place of earlier used refrigerants.

Table: Table showing various refrigerants used in food industries.

| Type of industry | Earlier refrigerant used | GWP | Alternative refrigerant used | GWP |
|---|---|---|---|---|
| Small scale | R134a | 1430 | R744 | 1 |
| Small scale | R22 | 1810 | R1234yf/R1234ze | 4-6 |
| Small scale | R404a | 3920 | Low GWP blends | <1300 |
| Large scale | R404a | 3920 | R717 | 0 |
| Large scale | R507a | 3990 | R744 | 1 |

Earlier, the use of chlorofluorocarbons (CFCs) and hydrochlorofluorocarbons (HCFCs) were in vogue for a long time due to their excellent thermodynamic properties. But as per recommendations of the Montreal and Kyoto protocols, they were being phased out due to their significant ODP and GWP values. As an immediate replacement of CFC-12, R-134a was adopted due to its low ODP value. Afterwards, alternative for R-134a was also needed to overcome the high GWP of this refrigerant. Figure shows the relative performances of other 12 alternative refrigerants with respect to R134a in terms of COP, volumetric displacement, mean temperature difference between the air and the refrigerant in the evaporator and compressor discharge temperature. The compositions of 12 alternative refrigerants have also been presented in table.

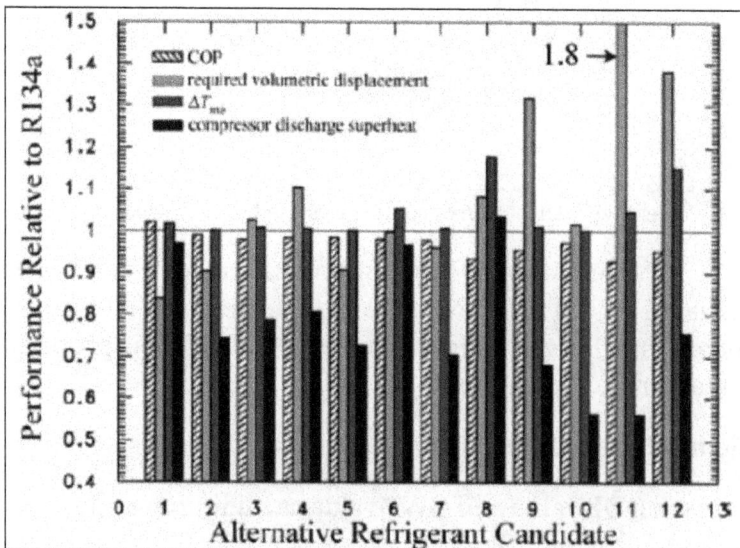

Showing the performance of alternative refrigerants.

Table: Table showing compositions of 12 alternative refrigerants.

| Alternative Refrigerant | Composition (mass %) |
|---|---|
| 1 | R134a/R1234yf/R32 (63/31/6) |
| 2 | R1234yf/R134a (56/44) |
| 3 | R134a/R1234ze/R134yf (42/40/18) |
| 4 | R1234ze/R134a (58/42) |
| 5 | R1234yf/R134a (60/40) |
| 6 | R1234ze/RR134a/R32 (53/40/7) |
| 7 | R1234yf/R152a/R134a (82/11/7) |
| 8 | R1234ze/R32/R152a (83/12/5) |
| 9 | R1234ze (100) |
| 10 | R1234yf (100) |
| 11 | 600a (100) |
| 12 | R600a/R290 (60/40) |

Different types of refrigeration systems implement different types of refrigerants by taking their thermodynamic properties into consideration. It has been observed that the usage of R404A is used in most of the sectors due to its excellent food conservation and freezing applications and widespread use in split type and centralised refrigeration systems. Figure shows the percentages of different refrigerants used in different food processing sectors.

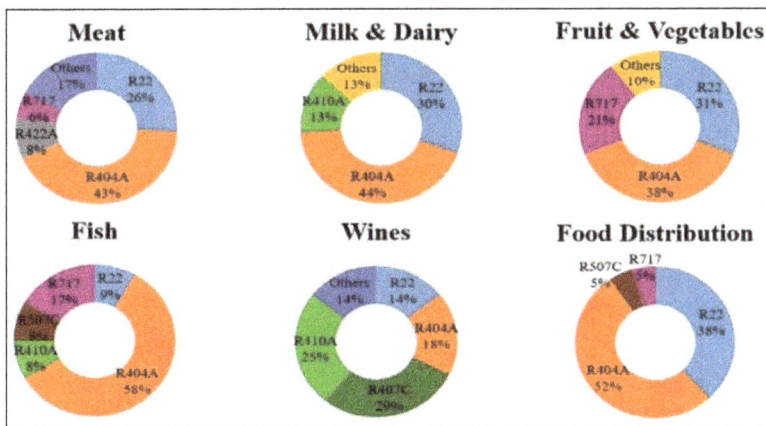

Figure: Various refrigerants used in different food processing sectors.

## Cold Chain

The term cold chain refers to a temperature-controlled supply chain that a refrigerated product passes through, which is then either until it is removed by a customer in a retail environment or unloaded from a delivery vehicle in the vicinity of its destination. For consumers, the cold chain is often related to transport, retail and household refrigerators.

But refrigeration is also used in the agri-food industry for the storage of raw materials and final products, as well as for food processing. An efficient and effective cold chain is designated to provide the best conditions for inhibiting any undesirable changes for as long as is practical. Effective refrigeration produces safe food with a long and quality life. A schematic diagram of a simple food chain has been presented in figure.

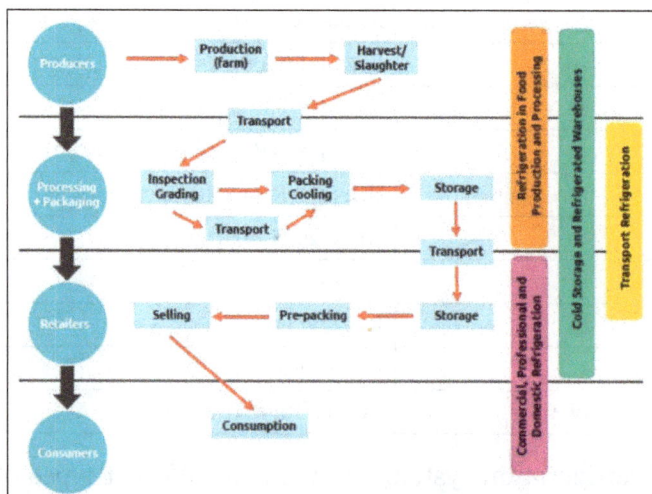

Overall cold chain and its components.

As food moves along the cold chain, it becomes a herculean task to control and regulate its temperature. This is because the temperatures of bulk packs of refrigerated products in large storerooms are far less sensitive to small heat inputs than single consumer packs in open display cases or in a domestic refrigerator.

## Refrigeration Cycle in Food Processing Industry

In food processing industries, the refrigeration systems in general operate with single-stage, direct expansion vapour compression refrigeration (VCR) cycle. These systems are very simple, reliable and also cheap. A VCR cycle consists of four basic components: compressor, evaporator, expansion valve and a condenser. A low-pressure cold liquid refrigerant is allowed to evaporate to a gas within the evaporator coil. This process requires heat, which is extracted, thus, cooling any medium surrounding the evaporator. The low-pressure hot gas from the evaporator is compressed in the compressor to a high-pressure hot gas. This high-pressure hot gas is then passed through another coil, where it condenses back to a high-pressure cold liquid. This process releases heat into any medium surrounding the condenser coil. This high-pressure cold liquid refrigerant then passes through the expansion valve where throttling process takes place, to a lower pressure section. It may be noted that the throttling process occurring in the expansion device is irreversible in nature. The low-pressure liquid then passes back to the evaporator. The schematic diagram of a simple mechanical VCR system and the corresponding T-s diagram have been shown in figure respectively. In general, factors influencing the nutrient content of refrigerated food include storage temperature, storage length, humidity and light.

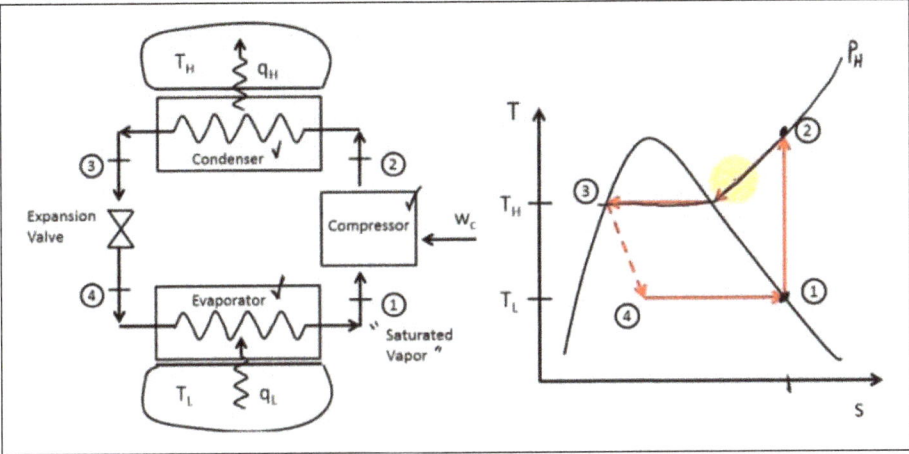

(a) Schematic and (b) T-s diagram of a VCR system.

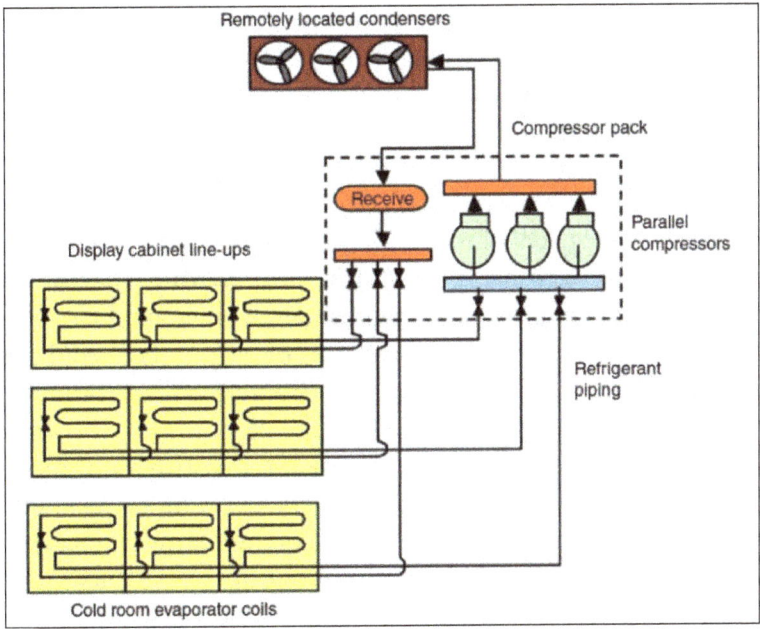

Schematic diagram of a centralized DX system.

The different processes undergoing in a VCR cycle with respect to the T-s diagram are:

Process 1-2: Isentropic work input to the compressor.

Process 2-3: De-superheating and condensation at constant pressure.

Process 3-4: Throttling in the expansion valve. As throttling process is irreversible in nature, it is represented by dotted line in the T-s diagram.

Process 4-1: Evaporation at constant pressure.

However, such systems have certain specifications in terms of the method for processing and the type of components used. For chilling, single-stage systems with flooded evaporators and screw compressors are recommended. Secondary refrigerants such as propylene glycol are preferred over to primary refrigerants as it significantly reduces the quantity of ammonia and promotes safety. For freezing, two-stage refrigeration systems with flooded evaporators and either of the piston or screw compressor is recommended. Refrigerants such as R134a or R717 are used in the high-temperature stage and R744 is used in the low temperature stage. As mentioned earlier, in most of the food processing industries, a centralised direct expansion (DX) system is established. Centralised systems provide the flexibility of installing the compressors and condensers in a centralised plant area, usually in the vicinity of any store. In the plant room, multiple refrigeration compressors, using common suction and discharge manifolds, are mounted on bases or racks normally known as compressor packs or racks which also contain all the necessary piping, valves, and electrical components needed for the operation and control of the compressors. The evaporators in the cold rooms are fed with refrigerant from the central plant through distribution pipework installed under the floor or along the ceiling of the sales area. Air-cooled or evaporative-cooling condensers used in conjunction with the multiple compressor systems are installed remotely from the compressors, usually on the roof of the plant room. Besides, separate compressor packs are used in chilling and freezing processes. The schematic diagram of a centralised DX system has been shown below in figure. However, the only major disadvantage associated with a centralised DX system is that a large amount of refrigerant is required, nearly 4-5 kg/kW of refrigeration capacity. To overcome such difficulty, a secondary or indirect system arrangement is used. With this arrangement, shown schematically in figure, a primary system can be located in a plant room or the roof and can use natural refrigerants such as hydrocarbons or ammonia to cool a secondary fluid, which is circulated to the coils in the display cabinets and cold rooms. Separate refrigeration systems and brine loops are used for the medium and low-temperature display cabinets and other refrigerated fixtures. The temperature of individual consumer packs, small individual items and especially thin sliced products responds very quickly to small amounts of added heat. All these products are commonly found in retail display cabinets and marketing constraints require that they have maximum visibility.

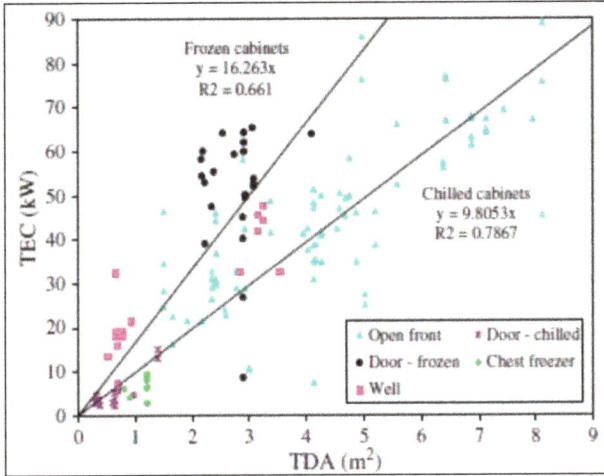

Relationship between TDA and TEC for all chilled and frozen cabinets.

## Refrigeration System Performance in Food Processing Unit

To understand how efficiently a system works, it is very important to know the performance of a system. COP is one of the main performance parameters in a refrigeration system. The COP of the system can be defined as the ratio of refrigerating effect produced to the power input to the system. Evans expressed the power input in terms of total energy consumption (TEC) and total display area (TDA). TEC can be expressed by the following equation,

TEC = DEC + REC

Where, DEC is direct electrical energy consumption, i.e. energy consumed by fans, heaters, defrost heaters, lighting, accessories and REC is the refrigeration electrical energy consumption and is calculated as:

$$REC = \frac{t_r \phi_0 \times (T_c - T_0)}{0.34 \times T_0}$$

Where, tr is defrost period in hr, Øo is the heat extraction rate in kilowatt based on EN441-12, Tc is the conventional condenser temperature at 308.18 K (35°C) for European comparisons, To is the refrigerant evaporator temperature in Kelvin based on EN44112. and 0.34 is taken to be the Carnot efficiency of refrigeration systems used in commercial refrigeration.

TDA can be evaluated by the following equation,

$$TDA = H_0 L_{oh} + H_g T_{gh} L_{gh} + V_g T_{gv} L_{gv}$$

Where, Ho is the open and Hg is the glazed horizontal projection (m), Vo is the open and Vg is the glazed vertical projection (m), Tgh is the light transmission through glazing surface for horizontal projection (%), Tgv is the light transmission through glazing surface for vertical projection (%), L is the cabinet length (m), Loh is the horizontal open length (m), Lov is the vertical open length (m), Lgh is the horizontal glazing length (m), and Lgv is the vertical glazing length (m).

Refrigeration techniques are vastly implemented in food processing industries for the storage, preservation and nourishment of food. For the storage, maintenance, transportation and other industrial parameters associated with food processing, a cold chain has been established. Industrial refrigeration is the first step in the cold chain where food is processed and stored before transport, retail and consumption. Performance of the system is analysed in terms of total energy consumption and total display area. The refrigeration sector faces many challenges with respect to reliability, energy consumption, environmental impact regulations and other economic concerns. Solutions for providing sustainable industrial refrigeration systems depend on the size of the facility and on the required temperature levels.

# Cooling

Cooling can be defined as a processing technique that is used to reduce the temperature of the food from one processing temperature to another or to a required storage temperature. Chilling is a processing technique in which the temperature of a food is reduced and kept at a temperature between −1°C and 8°C. The objective of cooling and chilling is to reduce the rate of biochemical and microbiological changes in foods, in order to extend the shelf-life of fresh and processed foods, or to maintain a certain temperature in a food process, e.g. in the fermentation and treatment of beer. Cooling is also used to promote a change of state of aggregation, e.g. crystallisation. In the wine industry, cooling (chilling) is applied to clarify the must before fermentation. The objective of cold stabilisation is to obtain the precipitation of tartrates (in wines) or fatty acids (in spirits) before bottling.

## Field of Application

Cooling is a process step in many food production processes. Chilling for food preservation is widely applied for a lot of perishable foods. The main application of cold stabilisation in the food industry is in the wine and spirit sector.

The supply of chilled foods to consumers requires a sophisticated distribution system, involving chilled stores, refrigerated transport and chilled retail display cabinets. Chilled foods can be grouped into three categories according to the storage temperature and the forth is applied in wine making:

- −1 °C to + 1 °C (fresh fish, meats, sausages and ground meats, smoked meats and fish).

- 0 °C to + 5 °C (pasteurised canned meat, milk and milk products, prepared salads, baked goods, pizzas, unbaked dough and pastry)

- 0 °C to + 8 °C (fully cooked meats and fish pies, cooked or uncooked cured meats, butter, margarine, cheese and soft fruits).

- 8 °C to 12 °C in the wine industry. The must is kept at this temperature between 6 and 24 hours.

## Description of Techniques, Methods and Equipment

Cooling of liquid foods is commonly carried out by passing the product through a heat exchanger (cooler) or by cooling the vessels. The cooling medium in the cooler can be groundwater, water recirculating over a cooling tower, or water (eventually mixed with agents like glycol) which is recirculated via a mechanical refrigeration system (ice-water). Cooling and chilling of solid foods is carried out by contacting the food with cold air, or directly with a refrigerant like liquid carbon dioxide or liquid nitrogen. The equipment used for freezing can also be used for cooling and chilling.

Some typical applications are given below:

## Cooling of Sugar

Sugar to be stored in silos must be dedusted and cooled to the storage temperature. This is done in a sugar cooler, which is a device in which warm and dried sugar is intensively aerated by cold filtered external air to cool the sugar to the storage temperature, approximately 20–30°C. The most common systems in use are coolers (typically drum or fluidised-bed coolers) with chilling systems with countercurrent or cross-current phase flow.

## Cryogenic Cooling

In cryogenic cooling the food is in direct contact with the refrigerant, which may be solid or liquid carbon dioxide, or liquid nitrogen. As the refrigerant evaporates or sublimates it removes heat from the food, thereby causing rapid cooling. Both liquid nitrogen and carbon dioxide refrigerants are colourless, odourless and inert.

## Cold Stabilisation

Cold stabilisation is a technique of chilling wines before bottling to cause the precipitation of harmless tartrate crystals. For spirits, this technique consists of bringing the spirit to a temperature of between -1°C and -7°C, depending on the operators, and possibly performing a stabulation (storing at low temperature) in a tank at constant temperature for between 24 and 48 hours. A cold filtration (around -1°C) allows the fatty acid esters to be retained.

For wines, three techniques can be employed: -stabilisation by batch and stabulation. This is the oldest technique and consists of bringing the wine to a temperature below zero close to the freezing point, then stabulating in an isothermal tank during a period of 5 to 8 days. But currently the most widely-used techniques are: -continuous stabilisation, where the stabulation tank is

replaced by a cylindro-conical crystalliser and an agitator, in which the wine will remain for only between 30 and 90 minutes, stabilisation by crystal seeding consisting of refrigerating at between -1° and -2°C, and seeding at 4 g/l of tartaric crystals with agitation over 2 to 4 hours, and later storage in tank, and decantation after 12 to 48 hours. There can be many variations on these basic schemes.

# Freezing

Freezing is one of the most widespread industrial methods of food preservation today. The transition from chilling to freezing is not merely a continuous change that can be explained on grounds of the lower temperature alone. On the contrary, freezing represents a point of sharp discontinuity in the relationship between temperature and the stability and sensory properties of foods.

- The exceptional efficiency of freezing as a method of food preservation is, to a large extent, due to the depression of water activity. Indeed, when food is frozen, water separates as ice crystals and the remaining non-frozen portion becomes more concentrated in solutes. This 'freeze concentration' effect results in the depression of water activity. In this respect, freezing can be compared to concentration and drying.

- On the other hand, the same phenomenon of 'freeze concentration' may accelerate reactions, inducing irreversible changes such as protein denaturation, accelerated oxidation of lipids and destruction of the colloidal structure (gels, emulsions) of the food.

- The rate of freezing has an important effect on the quality of frozen foods. Physical changes, such as the formation of large ice crystals with sharp edges, expansion, disruption of the osmotic equilibrium between the cells and their surroundings, may induce irreversible damage to the texture of vegetables, fruits and muscle foods. It has been established that such damage is minimized in the case of quick freezing.

## Phase Transition, Freezing Point

Figure shows 'cooling curves' describing the change in temperature as heat is removed from a sample. The graph at the left describes the cooling behavior of pure water. As the sample is cooled, the temperature drops linearly (constant specific heat) until the first crystal of ice is formed. The temperature at that moment is the ' freezing tempera- ture 'of pure water, o °C at atmospheric pressure, by definition. The freezing point is the temperature at which the vapor pressure of the liquid is equal to that of the solid crystal. Under certain conditions (absence of solid particles, slow undisturbed cooling), the sample may undergo a metastable state of supercooling. Some proteins, known as 'antifreeze proteins' are capable of preventing crystallization of water ice at the freezing point.

The graph at the right represents the cooling curve of a solution or a real food ma- terial. As the sample is cooled, the temperature drops linearly. The first ice crystal appears at the temperature $T_f'$. This is the temperature at which the water vapor pressure of the solution is equal to that of pure water ice. Since the water vapor pressure of a solution is lower than that of water at the same temperature, T &f is lower than the freezing temperature of pure water. The difference, called the 'freezing point depression ', increases with the molar concentration of the solution. Transformation of some of the liquid water to ice results in higher concentration of the solutes, which in turn lowers the freezing point and so on. There is no sharp phase transition at a constant temperature but rather a gradual 'zone of freezing', starting at the temperature of initial freezing, $T_f'$.

Experimental data pertaining to the initial freezing temperatures are available in the lit- erature. Assuming ideal behavior, the initial freezing temperature can be estimated from food composition data. The initial freezing temperature of common fruits and vegetables falls between - 0.8 and - 2.8.

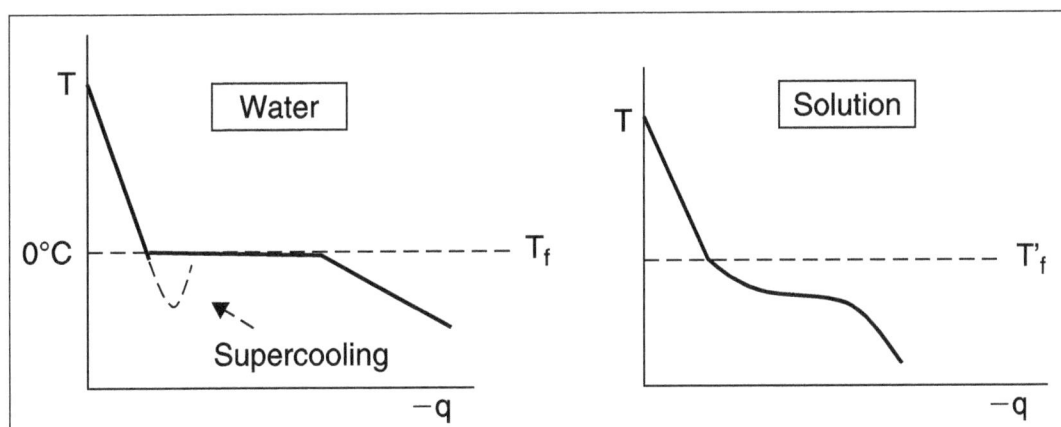

Cooling curves for pure water and aqueous solution.

Example:

Estimate the initial freezing temperature of 12 Bx orange juice and of 48 Bx orange juice concentrate. Assimilate the juices to 12% and 48% w/w solutions of glucose (MW = 180).

Solution:

The freezing point of a solution is the temperature at which its vapor pressure is equal to that of pure water ice. Assuming ideal behavior (Raoult's law), the vapor pressure of a solution is: $p = x_w \, p_o$, where $x_w$ is the molar fraction of water in the solution and $p_o$ is the vapor pressure of pure water. The molar fraction of the two solutions is:

For 12 Bx:

$$x_w = \frac{(100-12)/18}{(100-12)/18+12/180} = 0.987$$

For 48 Bx:

$$x_w = \frac{(100-48)/18}{(100-48)/18+48/180} = 0.915$$

Values of vapor pressure of water and ice at different temperatures are given in table. The vapor pressure of each solution at different temperatures is calculated from these values and the temperature at which they equal the vapor pressure of ice is searched. A graphical solution is shown in figure.

Results: The freezing point of 12 Bx juice is approximately - 1.5 °C.

The freezing point of 48 Bx juice concentrate is approximately - 9.2 °C.

It is not possible to freeze out all the water of a solution. When the concentration of the solute in the non-frozen portion reaches a certain level, that entire portion solidifies as though it were a pure substance. This new solid phase is called 'eutecticum'. A theoretical phase diagram for a salt solution is shown in Figure.

In the case of sugar solutions and most food materials, it is practically impossible to reach the eutectic point because glass transition of the non-frozen concentrated cold solution occurs before the eutectic point. Molecular motion in the glassy solid becomes extremely slow and any further crystallization of water ice becomes practically impossible.

## Freezing Kinetics and Freezing Time

The quality of frozen foods is strongly affected by the speed of freezing. Furthermore, freezing time is of obvious economic importance, as it determines the product throughput of the freezing equipment. An analysis of the factors that affect freezing time is, therefore, of interest.

Let us consider a mass of liquid, cooled by cold air blowing over its surface. We shall assume that the liquid is initially at its freezing temperature, so that all the heat removed from the liquid comes from the latent heat of freezing, released when some of the liquid is transformed to ice.

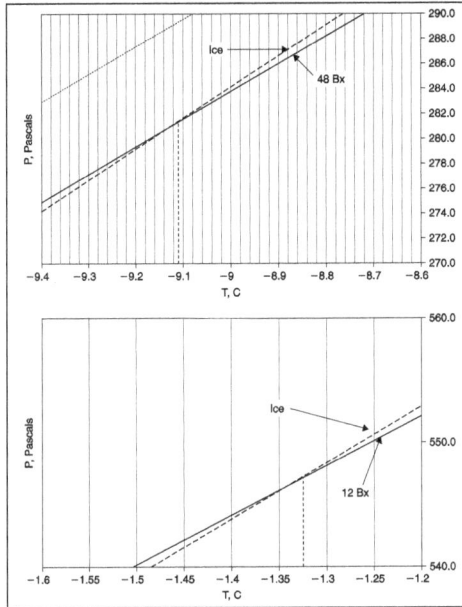

Freezing point of solutions.

The rate of heat transfer from the freezing front of the liquid to the cold air is therefore equal to the rate of heat release as a result of forming additional ice.

$$q = A\rho\lambda\frac{dz}{dt} = A\frac{1}{h+\dfrac{z}{k}}\left(T_f - T_a\right)$$

where:

q = rate of heat removal, W

A = area of heat transfer, m

$\rho$ = density of the liquid, kg . m$^{-3}$

$\lambda$ = latent heat of freezing of the liquid, J.kg$^{-1}$

z = thickness of the frozen phase

h = convective heat transfer coefficient at the air – ice interface, W. m$^{-2}$ .K$^{-1}$

k = thermal conductivity of the frozen phase, W. m$^{-1}$. K$^{-1}$

$T_a$ = temperature of the cooling medium (in this case, cold air)

$T_f$ = freezing temperature.

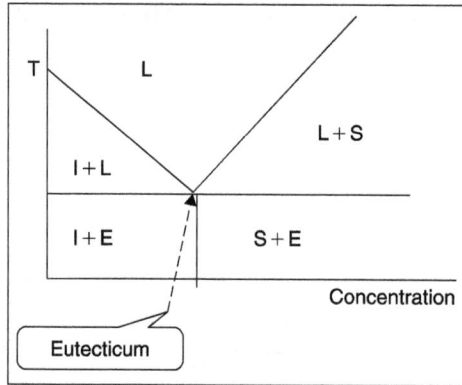

Phase diagram of a salt solution. S = salt; E = eutecticum; L = liquid; I = ice.

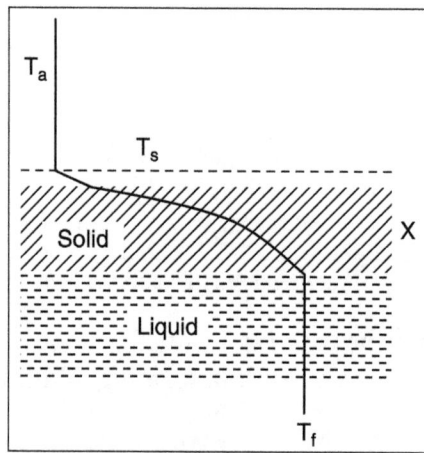

Temperature profile in freezing.

Rearranging and integrating we obtain the time necessary to freeze a mass of liquid of thickness Z:

$$\lambda = \frac{\rho \lambda}{T_f - T_a}\left(\frac{Z}{h} + \frac{Z^2}{2k}\right)$$

Equation $\lambda = \frac{\rho \lambda}{T_f - T_a}\left(\frac{Z}{h} + \frac{Z^2}{2k}\right)$ is known as the Plank equation, originally proposed by R.Z. Plank in 1913.

Equation $\lambda = \frac{\rho \lambda}{T_f - T_a}\left(\frac{Z}{h} + \frac{Z^2}{2k}\right)$ gives the freezing time for a semi-infinite body. For other geometries, Plank's formula is written in the following general form:

$$t = \frac{\rho \lambda}{T_f - T_a}\left(\frac{d}{Qh} + \frac{d^2}{Pk}\right)$$

For a plate of thickness d, cooled from both sides: $Q = 2\ P = 8$

For an infinite cylinder of diameter d: $Q = 4\ P = 16$

For a sphere of diameter d: $Q = 6\ P = 24$

Plank's formula is only approximate, due to the inaccuracy of some of the assumptions made:

- In reality, latent heat of freezing is not the only kind of energy exchanged. There are also some sensible heat effects such as the further cooling of the ice formed and lowering the temperature of the non-frozen material down to the freezing point. In practice, the error resulting from this assumption is not large, since the latent heat effects far exceed sensitive heat effects.

- As explained above, there is no sharp freezing point in foods. T f is therefore an average value.

- $\lambda$ refers to the latent heat of freezing of the food and not of pure water. If the mass fraction of water in the food is w and the latent heat of freezing of pure water is $\lambda_o$, then $\lambda$ can be calculated as $\lambda = w\ \lambda_o$. This is also an approximation since not all the water is freezable. Furthermore, if the food contains fats that are solidified in the process, the enthalpy of solidification should be taken into account in the calculation of $\lambda$.

More precise methods for the calculation of freezing time have been proposed. Some of these methods make use of numerical rather than analytical techniques. Despite its shortcomings, Plank's formula is valuable in process design and for visualizing the effect of process conditions on freezing time. Following are some of the practical consequences of Plank's equation:

1. Freezing time is inversely proportional to the overall temperature difference $T_f - T_a$.

2. Freezing time increases with increasing water content of the food.

3. Freezing time is proportional to the sum of two terms: a convective term proportional to the size d and a conductive term proportional to $d^2$. When freezing larger items, such as beef carcasses, whole birds or cakes, the conductive term (the internal resistance to heat transfer) becomes the predominant factor and the convective term becomes less significant, i.e. increasing the velocity is practically ineffective. On the other hand, when freezing small particles, the convective term is important and improving convective heat transfer to the surface (e.g. increasing the turbulence) results in considerably faster freezing.

Example:

Blocks of filleted fish, 5 cm thick are to be frozen in a plate freezer (by contact with a cold surface on both sides).

a. Estimate the time for complete freezing of the blocks. Data:

Temperature of the plates: − 28°C, constant

Average freezing temperature: − 5°C

Density of the fish: 1100 kg/m³

Water content of the fish: 70% w/w

Thermal conductivity of frozen fish : 1.7 W/m . K

Latent heat of freezing of water: 334 kJ/kg.

Sensible heat effects and heat loss will be neglected. Assume perfect contact between the fish blocks and the plate surface.

b. What will be the freezing time if the blocks are packaged in carton boxes? The thickness of the carton is 1.2 mm and its thermal conductivity is 0.08 W/mK.

Solution:

a. Equation $t = \dfrac{\rho\lambda}{T_f - T_a}\left(\dfrac{d}{Qh} + \dfrac{d^2}{Pk}\right)$ is applied, with Q = 2 and P = 8 (plate cooled from both sides).

$$t = \frac{\rho\lambda}{T_f - T_a}\left(\frac{d}{2h} + \frac{d^2}{8k}\right)$$

The surface contact is ideal, hence h is infinite. Substituting the data:

$$t = \frac{1100\times334000\times0.7}{-5-(-28)}\left(\frac{(0.05)^2}{8\times1.7}\right) = 2055s = 0.57h$$

The freezing time for the unpackaged fish is 0.57 hours.

b. The surface resistance to heat transfer is now not nil but equal to the thermal resistance of the package of thickness z:

$$\frac{1}{h} = \frac{z}{k} = \frac{1.2\times10^{-3}}{0.08} = 0.015 \Rightarrow h = 66.7$$

$$t = \frac{1100\times334000\times0.7}{-5-(-28)}\left(\frac{0.05}{66.7} + \frac{(0.05)^2}{8\times1.7}\right) = 10444s = 2.9h$$

The freezing time for the packaged fish is 2.9 hours.

Example:

a. Meat balls, 4 cm in diameter are to be frozen in a blast of cold air, at − 40°C and at 'moderate' velocity. The convective heat transfer coefficient at the air–meat ball interface is h = 10 W/m² K (determined from previous experiments with copper spheres). All other data are like for fish. Calculate the time for complete freezing.

b. Repeat the calculations for a 'very moderate' air current (h = 1 W/m² K), and a 'very turbulent' blast (h = 100 W/m² K).

c. Repeat the calculations for the 3 cases, but with 'mini' meat balls, 1 cm in diameter.

Solution:

a. Plank's equation for spheres is:

$$t = \frac{\rho\lambda}{T_f - T_a}\left(\frac{d}{6h} + \frac{d^2}{24k}\right)$$

Substitution of the data yields:

$$t = \frac{1100 \times 334000 \times 0.7}{-5 - (-28)}\left(\frac{0.04}{6 \times 10} + \frac{(0.04)^2}{24 \times 1.7}\right) = (1118174)(0.000706)$$

$$= 789s.$$

b. For very slow air (h = 1 W/ m² K):

$$t = 1118174\left(\frac{0.04}{6 \times 1} + \frac{(0.04)^2}{24 \times 1.7}\right) = (1118174)(0.00671) = 7503s.$$

For very turbulent air (h = 100 W/ m² K):

$$t = 1118174\left(\frac{0.04}{6 \times 100} + \frac{(0.04)^2}{24 \times 1.7}\right) = (1118174)(0.00106) = 118.5s.$$

c. For the mini-meatballs (d = 1 cm):

At moderate air velocity:

$$t = 1118174\left(\frac{0.01}{6 \times 10} + \frac{(0.01)^2}{24 \times 1.7}\right) = (1118174)(0.00169) = 189s.$$

Table: Effect of particle size and turbulence on freezing rate.

| Diameter | Freezing time (s) | | |
|---|---|---|---|
| | Low h | Moderate h | High h |
| 1 cm | 1867 | 189 | 21.5 |
| 4 cm | 7503 | 789 | 118.5 |

At slow air velocity:

$$t = 1118174 \left( \frac{0.01}{6 \times 1} + \frac{(0.01)^2}{24 \times 1.7} \right) = (1118174)(0.00167) = 1867 s.$$

At high air velocity:

$$t = 1118174 \left( \frac{0.01}{6 \times 100} + \frac{(0.01)^2}{24 \times 1.7} \right) = (1118174)(0.0000192) = 21.5 s.$$

The results are summarized in Table

Conclusion: the effect of heat transfer at the surface (e.g. air turbulence in blast freezing) is much stronger in the case of small items where internal resistance to heat transfer is less significant.

## Effect of Freezing and Frozen Storage on Product Quality

For a very large number of food products, freezing represents the best preservation method with respect to food quality. The nutritional value, the flavor and color of foods are affected very slightly, if at all, by the process of freezing itself. The main quality factor that may be adversely affected by freezing is the texture. On the other hand, unless appropriate measures are taken, the deleterious effect of long-term frozen storage and of thawing on every aspect of product quality may be significant.

## Effect of Freezing on Texture

In a vegetal or animal tissue, the cells are surrounded by a medium known as the extracellular fluid. The extracellular fluid is less concentrated than the protoplasm inside the cells. The concentration difference results in a difference in osmotic pressure which is compensated by the tension of the cell wall. This phenomenon, known as turgor, is the reason for solid appearance of meat and the crispiness of fruits and vegetables. When heat is removed from the food in the course of freezing, the extracellular fluid, being less concentrated, starts to freeze first. Its concentration rises and the osmotic balance is disrupted. Fluid flows from the cell to the extracellular space. Turgor is lost and the tissue is softened. When the food is thawed, the liquid that was lost to the extracellular space is not reabsorbed into the cell, but is released as free juice in the case of fruits or 'drip' in the case of meat.

It is generally believed that freezing damage to the texture of cellular foods is greatly reduced by accelerating the rate of freezing. This is explained by the fact that the condition of osmotic imbalance, created at the onset of freezing, disappears when the entire mass is frozen. Furthermore, rapid freezing results in the formation of smaller ice crystals, presumably less harmful to the texture of cellular systems.

Another probable reason for the deterioration of the texture is the volumetric expansion caused by freezing. The specific volume of ice is 9% higher than that of pure water. Because cellular tissues are not homogeneous with respect to water content, parts of the tissue expand more than the others. This creates mechanical stress that may result in cracks. Obviously, this effect is particularly strong in foods with high water content such as cucumbers, lettuce and tomatoes. This kind of texture damage is partially prevented by adding solutes. Adding sugar to fruits and berries before freezing was a widespread practice before the development of ultra-rapid freezing methods.

The rate of freezing affects the size of ice crystals. Slow freezing produces large crystals. It has been suggested that large crystals with sharp edges may break cell walls and contribute to texture deterioration in cellular foods subjected to slow freezing.

There is no controversy as to the fact that slow freezing results in higher percentage of drip in meat and fish. It is also accepted that rapid freezing causes less damage to the texture of particularly fragile fruits. On the other hand, the general applicability of the theory stating the superiority of rapid freezing seems to be questionable. Nevertheless, quick freezing continues to be the practical objective of food freezing process design.

## Effect of Frozen Storage on Food Quality

Frozen storage, even at fairly low temperature, does not mean the absence of deteriorative processes. On the contrary, frozen foods may undergo profound quality changes during frozen storage. While the rate of reactions is generally (but not always) slower in frozen foods, the expected shelf life, and therefore the time available for the reactions to take place, is long. Some of the frequent types of deterioration in frozen foods are protein denaturation resulting in toughening of muscle foods, protein–lipid interaction, lipid oxidation and oxidative changes in general (e.g. loss of some vitamins and pigments).

Extensive studies on the effect of frozen storage on product quality were undertaken in the 1960s by researchers and the Western Research Laboratories of the US Department of Agriculture. A large number of commodities were tested for changes in chemical composition and sensory characteristics. A concept known as 'time–temperature–tolerance' (TTT) was developed. The studies showed a linear relationship between the storage temperature and the logarithm of storage time for equal reduction in quality (loss of a certain vitamin, loss of color or loss of organoleptic score). The logical conclusion of these studies could be that lower storage temperature always results in higher quality. This is not always the case, however. Reactions may be accelerated by the 'freeze concentration' effect more than they are attenuated by the lower temperature. In this case, the rate of deterioration (e.g. lipid oxidation) may increase as the storage temperature is lowered, passing through a maximum and then diminishing at very low temperatures.

Mass transfer phenomena during frozen storage (oxygen transfer, loss of moisture) may be a major cause of quality loss. The quality of packaging is, therefore, particularly important in frozen foods. The PPP (product–process–package) approach consists of paying attention of all three factors in evaluating and predicting the effect of frozen storage on product quality.

Another type of change in frozen foods during storage is the process of recrystallization. Smaller crystals are more soluble than large ones. Equally, small ice crystals have a lower melting point then large ones. Consequently, if the storage temperature undergoes fluctuations, small ice crystals may melt and then solidify on the larger crystals. This may be the reason why foods frozen rapidly and those frozen slowly sometimes show similar ice crystal size distribution after storage. Recrystallization is particularly objectionable in ice cream, where conversion of small ice crystals to large ones results in loss of the smooth, creamy texture. The remedy is, of course, avoiding temperature fluctuation during storage as much as possible.

## Freeze-Drying

Freeze-Drying is the process of dehydrating frozen foods under a vacuum so the moisture content changes directly from a solid to a gaseous form without having to undergo the intermediate liquid state through sublimation. In this process, freeze dried food maintains its original size and shape with a minimum of cell rupture. Removing moisture prevents a product from deteriorating at room temperature.

The process is used for drying and preserving a number of food products, including meats, vegetables, fruits, and instant coffee products.

The dried product will be the same size and shape as the original frozen material and will be found to have excellent stability and convenient reconstitution when placed in water. Freeze dried products will maintain nutrients, color, flavor, and texture often indistinguishable from the original product.

Depending on the product and the packaging environment, freeze dried foods are shelf-stable at room temperature for up to twenty-five years or more, if canned, and between 6 months to 3 years if stored in a poly-bag container, making it perfect for survival food or food storage as well as for commercial use.

The main determinant of degradation is the amount and type of fat content and the degree to which oxygen is kept away from the product.

FRESH            FROZEN            FREEZE-DRIED

## How Freeze-drying Works

Simply put, freeze-drying is the removal of water from a frozen product using a process called sublimation. Sublimation occurs when a frozen liquid transforms directly to a gaseous state without passing back through the liquid phase. The process of freeze-drying consists of three phases: prefreezing, primary drying, and secondary drying.

## Pre-freezing

Freeze dried food must first be prefrozen below its eutectic temperature, or simply put, freezing the materials (solute) that make up the food. Although a product may appear to be frozen because of all the ice that is present, in actuality it is not completely frozen until all of the solute is frozen as well.

## Primary Drying

After prefreezing, ice must be removed from the product through sublimation. This requires careful control of two parameters; temperature and pressure. The rate of sublimation depends on the difference in vapor pressure of the product compared to the vapor pressure of the ice collector. Molecules move from the higher pressure sample to the lower pressure sample. Since vapor pressure is related to temperature, it is also necessary for the product temperature to be warmer than the ice collector temperature.

## Secondary Drying

After primary drying, all ice has sublimated but some liquid is still present in the product. Continued drying is necessary to remove the remaining water. The process for removing this excess water is called isothermal desorption. The excess water is desorbed from the product by making the product temperature higher than the ambient temperature.

During the entire freeze-drying process, the exact freezing methodology and proper storage is very important.

## The Benefits of Freeze-dried Food

There are several benefits to freeze-dried foods over dehydrating:

- Freeze-dried food retains original characteristics of the product, including:
  - Color
  - Form
  - Size
  - Taste
  - Texture
  - Nutrients
- Reconstitutes to original state when placed in water.
- Shelf stable at room temperature – cold storage not required.
- The weight of the freeze-dried products is reduced by 70 to 90 percent, with no change in volume.

- The product is light weight and easy to handle.

- Shipping costs are reduced because of the light weight and lack of refrigeration.

- Low water activity virtually eliminates microbiological concerns.

- Offers highest quality in a dry product compared to other drying methods.

- Virtually any type of food or ingredient, whether solid or liquid, can be freeze-dried.

## Advantages of Freeze-drying

The modern plants provide high-quality products for customers as well as unrivalled financial and operational advantages for the company by eliminating product loss, reduced energy costs and maximizing plant reliability and ease of use. Other benefits are:

- The plants offer an advanced technology and efficient design to ensure the preservation of excellent quality in a wide range of food products, such as vegetables, temperate and tropical fruits, fish, meat, TV-dinners, coffee, flavour essences and several other products.

- The original flavour, proteins and vitamins are preserved. The products will retain their original shape, colour and texture.

- Re-hydration is rapid and complete.

- The process results in stable products with long shelf life.

- The products are durable at a wide range of temperatures, eliminating the need for complicated cold chain distribution systems.

- The low weight and easy handling of freeze dried products reduce shipping costs dramatically.

## Applications of Freeze-drying

Freeze-drying is a relatively expensive process. The equipment is about three times as expensive as the equipment used for other separation processes, and the high energy demands lead to high energy costs. Furthermore, freeze-drying also has a long process time, because the addition of too much heat to the material can cause melting or structural deformations.

- Therefore, freeze-drying is often reserved for materials that are heat-sensitive, such as proteins , enzymes , microorganisms , and blood plasma . The low operating temperature of the process leads to minimal damage of these heat-sensitive products.

- Freeze-drying is used to preserve food and make it very lightweight.

- The process has been popularized in the forms of freeze-dried ice cream, an example of astronaut food.

- It is also popular and convenient for hikers because the reduced weight allows them to carry more food and reconstitute it with available water.

- Instant coffee is sometimes freeze-dried, despite high costs of freeze-dryers. The coffee is often dried by vaporization in a hot air flow, or by projection on hot metallic plates.

- Freeze-dried fruit is used in some breakfast cereal.

- However, the freeze-drying process is used more commonly in the pharmaceutical industry.

- In bacteriology freeze-drying is used to conserve special strain.

## Primary Drying

- During the primary drying phase, the pressure is lowered (to the range of a few millibars), and enough heat is supplied to the material for the water to sublimate.

- The amount of heat necessary can be calculated using the sublimating molecules' latent heat of sublimation.

- In this initial drying phase, about 95% of the water in the material is sublimated.

- This phase may be slow (can be several days in the industry), because, if too much heat is added, the material's structure could be altered.

- In this phase, pressure is controlled through the application of partial vacuum.

- The vacuum speeds sublimation, making it useful as a deliberate drying process.

- Furthermore, a cold condenser chamber and/or condenser plates provide a surface(s) for the water vapour to re-solidify on.

- This condenser plays no role in keeping the material frozen; rather, it prevents water vapor from reaching the vacuum pump, which could degrade the pump's performance. Condenser temperatures are typically below −50°C.

- In this range of pressure, the heat is brought mainly by conduction or radiation; the convection effect can be considered as insignificant.

## Secondary Drying

- The secondary drying phase aims to remove unfrozen water molecules, since the ice is removed in the primary drying phase.

- This part of the freeze-drying process is governed by the material's adsorption isotherms.

- In this phase, the temperature is raised higher than in the primary drying phase, and can even be above 0°C, to break any physico-chemical interactions that have formed between the water molecules and the frozen material.

- Usually the pressure is also lowered in this stage to encourage desorption (typically in the range of microbars, or fractions of a pascal). However, there are products that benefit from increased pressure as well.

- After the freeze-drying process is complete, the vacuum is usually broken with an inert gas, such as nitrogen, before the material is sealed.

- At the end of the operation, the final residual water content in the product is around 1 to 4%, which is extremely low.

## Properties of Freeze-dried Products

- If a freeze-dried substance is sealed to prevent the re-absorption of moisture, the substance may be stored at room temperature without refrigeration, and be protected against spoilage for many years. Preservation is possible because the greatly reduced water content inhibits the action of microorganisms and enzymes that would normally spoil or degrade the substance.

- Freeze-drying also causes less damage to the substance than other dehydration methods using higher temperatures.

- Freeze-drying does not usually cause shrinkage or toughening of the material.

- In addition, flavours and smells generally remain unchanged, making the process popular for preserving food. However, water is not the only chemical capable of sublimation, and the loss of other volatile compounds such as acetic acid (vinegar) and alcohols can yield undesirable results.

Freeze concentration is an excellent alternative to evaporation and reverse osmosis for concentration of many liquid foods. Product quality is generally high since low temperatures are used and no vapor-liquid interface occurs. However, current commercial freeze concentration technology is not economically competitive with the more established alternatives. Freeze concentration is applied where focus is on aroma retention and high quality products. It is specially suited for heat sensitive products. Freeze concentration is used for coffee extracts, fruit juices, vinegar, beer, wine and practically any other aqueous solution.

## Process

- Freeze concentration has been practiced for centuries. In its earliest form it was as simple as leaving a barrel of liquid outside in the cold winter night. Water would crystallize and grow as a thick layer of ice along the inside walls of the barrel. The next day they would simply cut a hole through the ice shell and drain the now concentrated product. The water (now ice) was simply discarded.

- Understanding the principles by which ice crystals grow in fluid foods would aid in furthering freeze concentration technology. In particular, if optimal heat balance conditions can be maintained throughout the freeze concentration process, large, easily separated crystals can be grown in short times. Modern freeze concentration processes consist of a crystallization section, where part of the water is converted into solid ice crystals using a refrigeration system. The ice crystals are then separated by filters, centrifuges or using the wash columns.

- Now, through a process of innovative engineering, process simplification and component standardization, the patented technology has reduced both equipment costs and energy usage significantly making Freeze Concentration a practical option for the constantly growing number of applications throughout the food and drink sector.

## References

- Refrigeration-in-food-processing-cold-chain: coolingindia.in, Retrieved 22 April, 2019
- Cooling-chilling, processing-technology: hyfoma.com, Retrieved 02 May, 2019
- Cap-food-process-engineering-technology-guide: labgraos.com.br, Retrieved 23 June, 2019
- ASHRAE (2006). 2006 ASHRAE Refrigeration Handbook. ASHRAE, New York .Dosset, R.J. and Horan, T.J. ( 2001 ). Principles of Refrigeration, Prentice Hall , New Jersey

# Post-Processing Operations

Post-processing operations are used in improving the quality of food and extend shelf life of some processed foods. Different operations such as materials handling, packaging, sealing, storage, etc. are used in the post processing of food. The topics elaborated in this chapter will help in gaining a better perspective about these post-processing operations.

## Materials Handling

Materials handling is defined as the organized movement of a specific material from one place to another, at the right time, and in the right quantity. It may involve lifting, moving horizontally or vertically, and storing materials. It is an important factor in the smooth operation of any food processing plant and can greatly affect the quality of the product and the cost and profitability of the operation.

From the time that a material enters the factory gate, cost is added every time it is handled. Each movement also carries the risk of damaging the material. Material handling is so commonplace that it is frequently overlooked in cost calculations. Everybody knows that shipping should be included as a cost. Material handling and flow should be seen as an internal shipping operation that has an associated monetary value. The cost of materials handling, storage, and movement must be added to the final product cost.

In many operations, production has been streamlined to be as efficient as possible, yet little or nothing has been done to make materials handling more efficient. It is no wonder that it costs can be as much as 50% of the total manufacturing cost. Improper handling of raw materials can be a major factor in waste and product damage. Every time a material is moved, there is danger of bumping, dropping, and crushing. Unsafe operation of industrial trucks, conveyors, and carts has been responsible for many serious plant accidents.

It affects the costs of manufacturing and distribution and the selling price of all food products. Modern handling methods are directed at accomplishing movement and distribution with a minimum of labor, waste, and cost, within the shortest time and with maximum safety.

In the food industry, materials handling is concerned with the movement of raw material from a warehouse, supplier, or the receiving bay to the bulk storage area or to a processing line. During processing, materials need to be moved from one process to another. After processing, materials need to be moved from packaging to a warehouse or to dispatch.

Setting up a materials handling plan requires a fair amount of effort, and the expectancy is that it will yield concomitant benefits. The objectives of materials handling include decrease the handling cost by better utilizing labor, machines, and space, decrease the operational cost, reduce production or processing time, increase efficient use of storage space, keep material moving, thereby reducing the space occupied by in-production material, prevent handling-related injuries and accidents, improve product quality, and reduce material waste.

## Packaging

Food packaging is the enclosing of food to protect it from damage, contamination, spoilage, pest attacks, and tampering, during transport, storage, and retail sale. The package is often labeled with information such as amount of the contents, ingredients, nutritional content, cooking instructions (if relevant), and shelf life. The package needs to be designed and selected in such a manner that there are no adverse interactions between it and the food. Packaging types include bags, bottles, cans, cartons, and trays.

Automated palletizer of bread with industrial robots at a bakery in Germany.

### Functions of Food Packaging

Food packaging serves many important functions. They may be broken down as follows.

1.  Containment: For items that are granulated, paper-based packages are the best, with a sealing system to prevent infiltration of moisture into the product. Other products are packaged using metal cans, plastic bags and bottles, and glass containers. Another factor in containment is packaging durability—in other words, the packaged food has to survive transport from the food processing facility to the supermarket to the home for the consumer.

2. Protection: The packaging must protect the food from (a) biological agents such as rats, insects, and microbes; (b) mechanical damage such as product abrasion, compressive forces, and vibration; and (c) from chemical degradation such as oxidation, moisture transfer, and ultraviolet light.

3. Communication: Packaged food must be identified for consumer use, mainly with label text and graphics. It can also be done by using special shapes for the food package, such as the Coca-Cola bottle or the can of Spam. Other well-known food package shapes include potato chip bags and milk bottles. These packages also detail nutritional information, and whether they are packaged according to kosher or halal specifications. The label may also indicate whether it is safe to put the packaged food (such as a TV dinner) through a microwave process.

4. Environmental issues: To protect the environment, we must be willing to reuse or recycle the packaging or reduce the size of the packaging.

5. Package safety: Before using a particular type of package for food, researchers must ensure that it is safe to use that packaging for the food being considered, and that there are no adverse interactions between the package and the food. This includes any metal contamination issues from a can to the food product or any plastic contamination from a bottle to the food product.

6. Product access: The packaging must be such that the product is readily accessible when the consumer is ready to use it. For example, pour spouts on milk cartons can make it easy to dispense the milk.

## Food Packaging Types

The materials mentioned above can be fashioned into different types of food packages and containers. Examples are given.

| Packaging type | Type of container | Examples of foods packaged |
|---|---|---|
| Aseptic packages | Primary | Liquid whole eggs |
| Plastic trays | Primary | Portion of fish |
| Bags | Primary | Potato chips |
| Bottles | Primary | Bottle of a soft drink |
| Boxes | Secondary | Box of soft drink bottles |
| Cans | Primary | Can of tomato soup |
| Cartons | Primary | Carton of eggs |
| Flexible packaging | Primary | Bagged salad |
| Pallets | Tertiary | A series of boxes on a single pallet, to transport packaged food from the manufacturing plant to a distribution center. |
| Wrappers | Tertiary | Used to wrap the boxes on the pallet for transport. |

Primary packaging is the main packaging that holds the food that is being processed. Secondary packaging combines the primary packages into a single box. Tertiary packaging combines all of the secondary packages into one pallet.

## Special Techniques

- Vacuum packaging or inert atmosphere packaging: Oxygen in the air tends to reduce the shelf life of food by the process known as oxidation. To prevent this process, some foods are packaged at reduced pressure (partial vacuum) or using an inert gas (such as nitrogen) to replace oxygen.

- Bags-In-Boxes: These are used for soft drink syrups, other liquid products, and meat products.

- Wine box: This is a type of box designed for storage of wine.

## Packaging Machines

The design and use of packaging machinery needs to take into account the following factors: technical capabilities, labor requirements, worker safety, maintainability, serviceability, reliability, ability to integrate into the packaging line, capital cost, floorspace, flexibility of use, energy usage, quality of outgoing packages, qualifications (for food, phamaceuticals, and so forth), throughput, efficiency, productivity, and ergonomics.

Packaging machines may be of the following general types:

- Blister, Skin and Vacuum Packaging Machines;
- Capping, Over-Capping, Lidding, Closing, Seaming and Sealing Machines;
- Cartoning Machines;
- Case and Tray Forming, Packing, Unpacking, Closing and Sealing Machines;
- Check weighing machines;
- Cleaning, Sterilizing, Cooling and Drying Machines;
- Conveying, Accumulating, and Related Machines;
- Feeding, Orienting, Placing, and Related Machines;
- Filling Machines: handling liquid and powdered products;
- Package Filling and Closing Machines;
- Form, Fill and Seal Machines;
- Inspecting, Detecting and Checkweighing Machines;
- Palletizing, Depalletizing, Pallet Unitizing and Related Machines;

- Product Identification: for labeling, marking, and so forth;

- Wrapping Machines;

- Converting Machines;

- Other specialty machinery.

## Types of Containers

The technical purposes of packaging are:

- To contain foods (to hold the contents and keep them clean and secure without leakage or breakage until they are used).

- To protect foods against a range of hazards during distribution and storage (to provide a barrier to dirt, micro-organisms and other contaminants, and protection against damage caused by insects, birds and rodents, heat, oxidation, and moisture pickup or loss).

- To give convenient handling throughout the production, storage and distribution system, including easy opening, dispensing and re-sealing, and being suitable for easy disposal, recycling or re-use.

- To enable the consumer to identify the food, and give instructions so that the food is stored and used correctly.

Polypropylene drums used as shipping containers for fruit pulp.

The shelf life of a food is the length of time it can be stored before the quality becomes unacceptable, and this includes the time to distribute food to retailers and store it by the consumer. It is important to note that the selection of a packaging material for a particular food depends not only on its technical suitability (i.e. how well the package protects a food for the required shelf life), but also on the availability and cost in a particular area, and any marketing considerations that favour choosing a certain type of package.

Packaging is important because it aids food distribution, and rapid and reliable distribution helps remove local food surpluses, allows consumers more choice in the foods available and helps to reduce malnutrition. Packaging also reduces post harvest losses, which together with giving access to larger markets, allows producers to increase their incomes. Therefore, adequate packaging in developing countries has profound effects on both the pattern of food consumption and the amount of food consumed.

Packaging materials can be grouped into two main types:

- Shipping containers, which contain and protect the contents during transport and distribution but have no marketing function. Examples include sacks, corrugated fibreboard (cardboard) cartons, shrink-wrapped or stretch-wrapped containers, crates, barrels or drums.

- Retail containers, which protect and advertise the food for retail sale and home storage. Examples include metal cans, glass or plastic bottles and jars, plastic tubs, pots and trays, collapsible tubes, paperboard cartons and flexible plastic or paper bags, sachets and overwraps.

Frequently more than one type of material is used to package a single product. For example, display cartons may contain multiple packs of food packaged in flexible film. These in turn are placed in corrugated board shipping boxes and loaded onto pallets.

## Types of Packaging Materials - Traditional Materials

These materials have been used since the earliest times for domestic storage and local sales of foods. However, with the exception of glazed pottery, they have poor barrier properties and are only used to contain foods and keep them clean. They are also unsuited to the needs of commercial production processes and are considered by many customers to be less attractive than the newer 'industrial' materials described. A summary of the main types of traditional materials and possible current uses are as follows:

Traditional leaf packaging of coffee beans.

## Leaves, Vegetable Fibres and Textiles

Leaves are cheap and readily available, and are used as wrappers for products such as cooked foods that are quickly consumed. Banana or plantain leaves are used for wrapping traditional cheese and fruit confectionery such as guava cheese. Maize leaves are used to wrap corn paste or blocks of brown sugar, and 'Pan' leaves are used for wrapping spices in India. Other examples are green coconut palm, papyrus leaves and bamboo and rattan fibres, which are woven into bags or baskets and used for carrying meat and vegetables in many parts of the world. Palmyra palm leaves are used to weave boxes in which cooked foods are transported, and small banana leaf bags are used to contain coffee beans that are a traditional gift in some parts of Africa. Some of these have the potential to be developed as niche packaged products for tourist markets.

Fibres from kenaf and sisal plants are mainly used for making ropes, cord and string, which can be made into net bags to transport hard fruits. They can also be spun into a yarn that is fine enough to make coarse canvas. Other examples of textile containers are woven jute sacks, which are used to transport a wide variety of bulk foods including grain, flour, sugar and salt. Plant fibre sacks are flexible, lightweight and resistant to tearing, have good durability, and may be chemically treated to prevent them rotting. Their rough surfaces are non-slip, which makes stacking easier compared to synthetic fibre sacks, and they are bio-degradable. Most textile sacks can be re-used several times after washing and they are easily marked to indicate the contents. They are still widely used to transport fresh or dried crops, but they are being replaced as shipping containers by woven polypropylene or multi-walled paper sacks. Calico is a closely woven, strong cotton fabric that can be made into bags for flour, grains, legumes, coffee beans and sugar. Muslin and cheesecloth are open-mesh, light fabrics used both to strain liquid foods during processing and to wrap foods such as cheeses and processed meats (e.g. smoked ham).

## Wood

Wooden containers protect foods against crushing, have good stacking characteristics and a good weight-to-strength ratio. Wooden boxes, trays and crates have traditionally been used as shipping containers for a wide variety of solid foods including fruits, vegetables and bakery products. Wooden tea chests are produced more cheaply than other containers in tea-producing countries and are still widely used. However, in most applications, plastic containers have a lower cost, are more easily cleaned for reuse, do not risk contaminating foods with splinters, and they have largely replaced wooden containers in most applications. Small wooden boxes are used to pack tea or spices for tourist markets in some countries. Wooden barrels have been traditionally used as shipping containers for a wide range of liquid foods, including cooking oils, wine, beer and juices. They continue to be used for some wines and spirits because flavour compounds from the wood improve the quality of the products, but in other applications have been replaced by aluminium, coated steel or plastic barrels.

## Leather

Leather containers made from camel, pig or kid goat hides have traditionally been used as flexible, lightweight, non-breakable containers for water, milk and wine. Manioc flour and solidified sugar were also packed in leather cases and pouches, but the use of leather has now ceased for most commercial food applications.

## Earthenware

Pottery is still used domestically for storage of liquid and solid foods such as yoghurt, beer, dried foods, honey, etc. Corks, wooden lids, wax or plastic stoppers, or combinations of these are used to seal the pots. If they are glazed and well sealed, they prevent oxygen, moisture and light from entering the food and they are therefore suitable for storing oils and wines. They also restrict contamination by micro-organisms, insects and rodents. Unglazed earthenware bowls or pots are porous and the evaporating moisture makes them suitable for products that need cooling. They are still used for local sales of curd or yoghurt in parts of Asia.

Glass or plastic containers have largely replaced pottery because of its high weight, fragility, variability in volume when hand-made, and the difficulty of adequately cleaning pottery containers for re-use.

## Types of Packaging Materials - Industrial Materials

These materials have been developed over the last 200-300 years and are the main types of packaging used by small-scale food processors. The availability of glass, metal or plastic containers varies considerably in different countries, and this, together with the relative cost of different materials, determines their uptake by local processing industries. Where these materials have to be imported, large minimum order sizes can be a significant constraint on the development of a processing sector.

## Metal Containers

There are two basic types of metal cans: those that are sealed using a 'double seam' and are used to make canned foods; and those that have push-on lids or screw-caps that are used to pack dried foods (e.g. milk or coffee powder, dried yeast) or cooking oils respectively. Double-seamed cans are made from tinplated steel or aluminium and are lined with specific lacquers for different types of food. Cans have a number of advantages over other types of container: when sealed with a double-seam they provide total protection of the contents; they are tamperproof; and they can be made in a wide range of shapes and sizes. However, the high cost of metal and the high manufacturing costs make cans expensive compared to other containers. They are heavier than plastic containers and therefore have higher transport costs. There are few can-making factories in developing countries and small-scale food processors generally do not use metal cans because of these disadvantages and/

or lack of availability. Larger (200 litre) metal drums are widely used as shipping containers for oils, juices and other liquid foods, although cheaper plastic drums are steadily replacing them. Other types of metal containers include aluminium foil cups and trays, laminated foil pouches as alternatives to cans or jars, collapsible aluminium tubes for pastes, and aluminium barrels. The advantages of aluminium are that it is impermeable to moisture, odours, light and microorganisms, and is an excellent barrier to gases. It has a good weight:strength ratio and a high quality surface for decorating or printing.

## Glass

Glass bottles and jars have some of the advantages of metal cans: they are impervious to microorganisms, pests, moisture, oxygen and odours; they do not react foods or have chemicals that migrate into foods; they can be heat processed; they are recyclable, and (with new lids) they are re-useable; they are rigid, to allow stacking without damage; and unlike metal cans, they are transparent to display the contents. The main disadvantages of glass are: the higher weight than most other types of packaging, which incurs higher transport costs; containers are easily broken, especially when transported over rough roads; they have more variable dimensions than metal or plastic containers; and there are potentially serious hazards from glass splinters or fragments that can contaminate foods. Glass containers are still widely used for foods such as juices, wines, beers, pickles/chutneys and jams, especially in countries that have a glass-making factory, but their disadvantages and the high cost for imported containers in other places mean that they are steadily being replaced by plastic containers.

Glass jars used by a smallscale jam maker.

# Paper and Cardboard

Paper and boards are made from wood pulp and additives are mixed into the pulp to give particular properties to the packaging, including:

- Fillers such as china clay, to increase the brightness of paper and improve surface smoothness and printability.

- Binders, including starches, vegetable gums, and synthetic resins to improve the strength.

- Resin or wax sizing agents to reduce penetration by water or printing inks.

- Pigments to colour the paper and other chemicals to assist in the manufacturing process.

Different types of paper are used to wrap foods: 'sulphate' paper is strong and used for single- or multi-walled paper sacks for flour, sugar, fruits and vegetables; 'Sulphite' paper is lighter and weaker and is used for grocery bags and sweet wrappers, as an inner liner for plastic biscuit wrappers and laminated with plastic films. Greaseproof paper is sulphite paper made resistant to oils and fats, and used to wrap meat and dairy products. 'Glassine' is a greaseproof sulphite paper that is given a high gloss to make it resistant to water when dry, but it loses its resistance once it becomes wet. Tissue paper is a soft paper used for example to protect fruits against dust and bruising. Papers are also treated with wax to provide a moisture barrier and allow the paper to be heat sealed. Wax coatings are easily damaged and the wax is therefore laminated between layers of paper and/or polyethylene when used for bread wrappers and inner liners for cereal cartons.

'Paperboard' is a term that includes boxboard, chipboard and corrugated or solid fibreboards. Typically, paperboard has the following structure:

- A top layer of white material to give surface strength and printability.

- Middle layers of grey/brown lower grade material.

- An under-layer of white material to stop the colour of the middle layer showing through.

- A back layer if strength or printability are required.

All layers are glued together with adhesive.

White board is suitable for contact with foods and is often coated with wax or laminated with plastic to make it heat sealable. It is used for ice cream, chocolate and frozen food cartons. Chipboard is made from recycled paper and is used for example as the outer cartons for tea or cereals but not in contact with foods. It may be lined with white board to improve the appearance and strength. Other types include moulded paperboard trays for eggs, fruit, meat or fish or for egg cartons.

Small paperboard tubs or cans are used for snackfoods, confectionery, nuts, salt, cocoa powder and spices. Larger drums are used as a cheaper alternative to metal drums for powders and other dry foods and, when lined with polyethylene, for cooking fats. They are lightweight, resist compression and may be made water resistant for outside storage. Other products that are handled in lined drums include fruit and vegetable products, peanut butter and sauces. Corrugated board resists impact, abrasion and compression damage, and is therefore used for shipping containers. Smaller more numerous corrugations give rigidity, whereas larger corrugations or double- and triple-wall corrugated material provides cushioning and resists impact damage. Corrugated cartons are used as shipping containers for bottled, canned or plastic-packaged foods. Wet foods may be packed by lining the corrugated board with polyethylene or a laminate of wax-coated greaseproof paper and polyethylene, and used for chilled bulk meat, dairy products and frozen foods.

## Flexible Plastic Films

In general, flexible plastic films have relatively low cost and good barrier properties against moisture and gases; they are heat sealable to prevent leakage of contents; they add little weight to the product and they fit closely to the shape of the food, thereby wasting little space during storage and distribution; they have wet and dry strength, and they are easy to handle and convenient for the manufacturer, retailer and consumer. The main disadvantages are that (except cellulose) they are produced from non-renewable oil reserves and are not biodegradable. Concern over the environmental effects of non-biodegradable oil-based plastic packaging materials has increased research into the development of 'bioplastics' that are derived from renewable sources, and are biodegradable. However, these materials are not yet available commercially in developing countries.

There is a very wide choice of plastic films made from different types of plastic polymer. Each can have ranges of mechanical, optical, thermal and moisture/gas barrier properties. These are produced by variations in film thickness and the amount and type of additives that are used in their production. Some films (e.g. polyester, polyethylene, polypropylene) can be 'oriented' by stretching the material to align the molecules in either one direction (uniaxial orientation) or two (biaxial orientation) to increase their strength, clarity, flexibility and moisture/gas barrier properties. There are thus a very large number of plastic films and small-scale processors should obtain professional advice when selecting a material to ensure that it is suitable for the intended product and shelf life. Typically, the information required includes: type of plastic polymer(s) required; thickness/strength; moisture and gas permeability; heat seal temperature; printability on one or both sides; and suitability for use on the intended filling machinery.

A summary of the main different types of flexible plastic films is as follows:

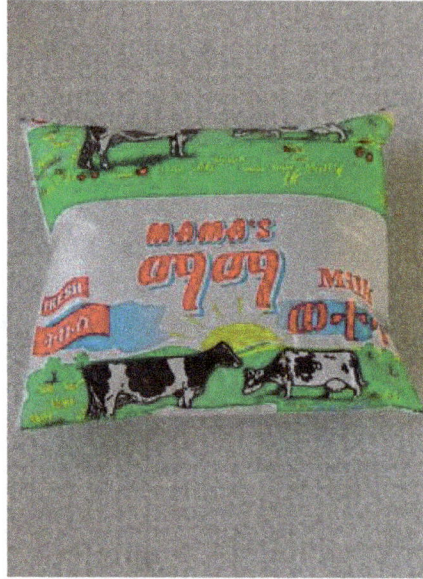

Milk packaged in flexible film.

## Cellulose

Plain cellulose is a glossy transparent film that is odourless, tasteless and biodegradable (within approximately 100 days). It is tough and puncture resistant, although it tears easily. It has dead-folding properties that make it suitable for twist-wrapping (e.g. sugar confectionery). However, it is not heat sealable and the dimensions and permeability of the film vary with changes in humidity. It is used for foods that do not require a complete moisture or gas barrier, including fresh bread and some types of sugar confectionery. Cellulose acetate is a clear, glossy transparent, sparkling film that is permeable to water vapour, odours and gases and is mainly used as a window material for paperboard cartons.

## Polyethylene

Low-density polyethylene (LDPE) is heat sealable, inert, odour free and shrinks when heated. It is a good moisture barrier but is relatively permeable to oxygen and is a poor odour barrier. It is less expensive than most films and is therefore widely used for bags, for coating papers or boards and as a component in laminates. LDPE is also used for shrink- or stretch-wrapping. Stretch-wrapping uses thinner LDPE (25 - 38 μm) than shrink-wrapping (45 - 75 μm), or alternatively, linear low-density polyethylene is used at thicknesses of 17 - 24 μm. The cling properties of both films are adjusted to increase adhesion between layers of the film and to reduce adhesion between adjacent packages.

High-density polyethylene (HDPE) is stronger, thicker, less flexible and more brittle than LDPE and a better barrier to gases and moisture. Sacks made from HDPE have high tear and puncture resistance and have good seal strength. They are waterproof and chemically resistant and are increasingly used instead of paper or sisal sacks.

## Polypropylene

Polypropylene is a clear glossy film with a high strength and puncture resistance. It has a moderate barrier to moisture, gases and odours, which is not affected by changes in humidity. It stretches, although less than polyethylene. It is used in similar applications to LDPE. Oriented polypropylene is a clear glossy film with good optical properties and a high tensile strength and puncture resistance. It has moderate permeability to gases and odours and a higher barrier to water vapour, which is not affected by changes in humidity. It is widely used to pack biscuits, snackfoods and dried foods.

## Other Films

Polyvinylidene chloride is very strong and is therefore used in thin films. It has a high barrier to gas and water vapour and is heat shrinkable and heat sealable. However, it has a brown tint which limits its use in some applications. Polyamides (or Nylons) are clear, strong films over a wide temperature range (from – 60 to 200°C) that have low permeability to gases and are greaseproof. However, the films are expensive to produce, require high temperatures to heat seal, and the permeability changes at different storage humidities. They are used with other polymers to make them heat sealable at lower temperatures and to improve the barrier properties, and are used to pack meats and cheeses.

## Coated Films

Films are coated with other polymers or aluminium to improve their barrier properties or to impart heat sealability. For example a nitrocellulose coating on both sides of cellulose film improves the barrier to oxygen, moisture and odours, and enables the film to be heat sealed when broad seals are used. Packs made from cellulose that has a coating of vinyl acetate are tough, stretchable and permeable to air, smoke and moisture. They are used for packaging meats before smoking and cooking. A thin coating of aluminium (termed 'metallisation') produces a very good barrier to oils, gases, moisture, odours and light. This metallised film is less expensive and more flexible than plastic/aluminium foil laminates. The properties of selected coated films are shown in Table.

Table: Properties of selected packaging materials.

| Film | Coating | Barrier to | | Strength | Clarity | Thickness ($\mu m$) |
|---|---|---|---|---|---|---|
| | | Moisture | Air/odours | | | |
| Cellulose | PvDC | * | *** | * | *** | 21-40 |
| | Aluminium | *** | *** | * | *** | 19-42 |
| | Nitro- cellu- lose | *** | *** | * | - | 21-42 |

| LDPE | - | ** | * | ** | * | 25-200 |
|---|---|---|---|---|---|---|
| HDPE | - | *** | ** | *** | * | 350-1000 |
| Polypropylene | - | *** | *** | *** | *** | 20-40 |
| | PvDC | *** | *** | *** | *** | 18-34 |
| | Aluminium | *** | *** | *** | - | 20-30 |
| Polyester | - | ** | ** | *** | ** | 12-23 |
| | - | *** | *** | *** | - | 20-30 |

* = low, ** = medium, *** = high. PvDC = polyvinylidene chloride, LPDE = low density polyethylene, HDPE = High density polyethylene.

Thicker films of each type have better barrier properties than thinner films.

## Laminated films

Lamination (bonding together) of two or more films improves the appearance, barrier properties or mechanical strength of a package. Some examples of laminated films are shown in table.

Table: Selected laminated films used for food packaging.

| Laminated film | Typical food applications |
|---|---|
| Polyvinylidene chloride coated polypropylene (2 layers) | Crisps, snackfoods, confectionery, ice cream, biscuits, chocolate |
| Polyvinylidene chloride coated polypropylene-polyethylene | Bakery products, cheeses, confectionery, dried fruit, frozen vegetables |
| Cellulose-polyethylene-cellulose | Pies, crusty bread, bacon, coffee, cooked meats, cheeses |
| Cellulose acetate-paper-foil-polyethylene | Dried soups |
| Metallised polyester-polyethylene | Coffee, dried milk |
| Polyethylene-aluminium-paper | Dried soup, dried vegetables, chocolate |

## Coextruded Films

Coextrusion is the simultaneous extrusion of two or more layers of different polymers to make a film. Coextruded films have three main advantages over other types of film: they have very high barrier properties, similar to laminates but produced at a lower cost; they are thinner than laminates and are therefore easier to use on filling equipment; and the layers do not separate. There are three main groups of polymers that are coextruded:

- Low-density and high-density polyethylene, and polypropylene.

- Polystyrene and acrylonitrile-butadiene-styrene.

- Polyvinyl chloride.

Typically a three-layer coextrusion has an outside layer that has a high gloss and print-ability, a middle bulk layer which provides stiffness and strength, and an inner layer which is suitable for heat sealing. They are used, for example, for confectionery, snack-foods, cereals and dried foods. Thicker coextrusions (75 - 3000 $\mu m$) are formed into pots, tubs or trays.

Table: Selected applications of coextruded plastic films.

| Coextruded film | Typical food applications |
|---|---|
| High impact polystyrene-PET | Margarine, butter tubs |
| Polystyrene- polystyrene-PvDC- polystyrene | Juice and milk bottles |
| Polystyrene- polystyrene-PvDC-polyethylene | Tubs for butter, cheese, margarine, bottles for coffee, mayonnaise, sauces. |

## Rigid and Semi-rigid Plastic Containers

There is a wide range of plastic bottles, pots, jars, trays and tubs made from single or coex-truded plastics that are increasingly used for processed foods, when they are available in developing countries. The main advantages, compared with glass and metal, are as follows:

- Lower weight, resulting in savings of up to 40% in transport and distribution costs. Cups, tubs and trays are tapered (a wider rim than base) for more compact stacking for transport and storage.

- Lower production costs using less energy.

- Tough, unbreakable and easy to seal.

- Very good barrier properties.

- Precisely moulded into a wider range of shapes than glass or metal containers.

- Can be coloured for consumer appeal and to give UV-light protection to foods. However, they are not re-usable, are less rigid than glass or metal for stacking and cannot be heated to the same high temperatures as glass and metal. They are used for example as:

  ○ Cups or tubs for margarine, processed meats, cheese, spreads, yoghurt, pea-nut butter, dried foods or ice cream and desserts (high-nitrile resin copoly-mers or high-impact polystyrene and acrylonitrile butadiene styrene).

  ○ Trays for meat products and chocolates, tubs for margarine or jams, and (polyvinyl chloride).

- good oil resistance and low gas permeability.

- Bottles and jars for fruit juices, squashes and juice concentrates, vinegar, cooking oil, milk, wine, syrup and, and as drums for salt and bulk fruit juices (HDPE, polyvinyl chloride).

- Bottles for carbonated drinks (polyethylene terephthalate (PET) - PET is a very strong transparent glossy film that is a good moisture and gas barrier. It is biaxially oriented to develop the strength for use in carbonated drinks bottles.

- Squeezable bottles and pots for mustard, mayonnaise, jams, tomato ketchup and other sauces (polypropylene coextruded with ethylene vinyl alcohol).

- Trays for chocolates, eggs, or soft fruit.

- Foam cartons or trays for eggs, fresh fruits and takeaway meals (polystyrene).

# Filling and Sealing Packaged Foods

Solid foods are either in the form of large pieces (e.g. cut fruit and fish) or particles that flow like liquids (e.g. powders, rice, beans, maize etc.). At a small scale of operation, large pieces are usually packed by hand whereas powders and small particulate foods can often be filled using similar fillers to those used for liquids.

Liquids can be either thin (e.g. milk, wines and juices) or thick (viscous) such as oils, pastes, creams, sauces or jams. No one type of filling machine is suitable for all types of foods and the selection of suitable equipment depends on the viscosity, temperature, particle size and foaming characteristics of the product, and the production rate required.

## Filling Thin Liquids and Particulate Foods

The simplest manual filler is a jug, but this is often too slow. A simple manual filling machine for liquids is made by fitting one or more taps to the base of a large bucket or tank. The bucket should be stainless steel for filling hot acidic liquids (e.g. fruit juices) or food-grade plastic for cold filling. The taps should be 'gate-valve' types and not domestic water taps, which are more difficult to clean. In manual filling, the amount of food dispensed into the container is judged by the operator, and training is required to ensure that consistent volumes are filled into every container.

There are a variety of dispensing machines that control the volume of liquid that is filled into each container, and do not rely on the judgement of an operator. Timed gravity fillers are an economical type of volumetric filling machine, but the range of applications is limited to low-viscosity liquids that do not foam (e.g. bottled water and alcoholic spirits). The product is contained in a tank above a set of pneumatically operated valves. Each valve is independently timed to deliver precise amounts of liquid under gravity into the containers. Another type of filler is a 'dispenser' that is fitted with a 3-way valve. In the first position the valve allows a cylindrical chamber to fill from a tank above. It is then moved to the second position to empty the food into a container below. The volume of food in the cylinder can be adjusted to fill different sized containers.

1. Filling cylinder from product tank.

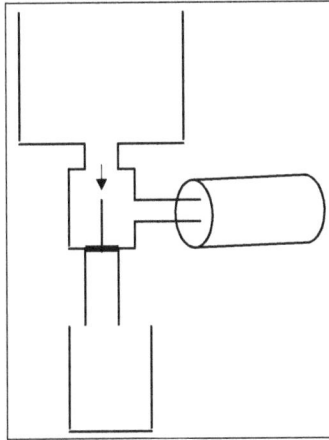

2. Emptying cylinder into container.

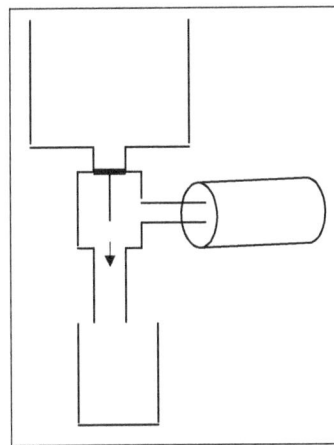

Operation of liquid dispenser.

Machines for filling powders and particulate foods have a hopper that is filled with food, and when the weight of food reaches a pre-set limit, the base of the hopper opens to drop the powder into a container below. The pre-set weight can be adjusted to fill different sized containers. These machines are used to fill flour, grains and other foods that have uniform sized pieces.

1. Filling the hopper.

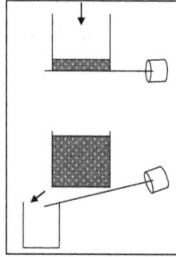

2. Discharging preset weight into container.

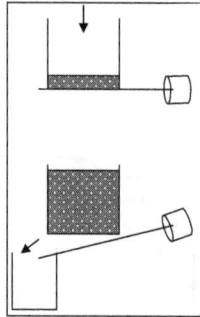

Filler for particulate foods.

## Filling Viscous Liquids and Liquids that contain Particles

Because viscous liquids do not flow easily, a dispenser or piston filler is the preferred option. Dispensers are similar in operation to the equipment described in figure, but have a moving piston in the cylinder to pump viscous products into the container). This equipment is relatively low-cost but has lower production rates than piston fillers. A small gear-pump filler can be adjusted to dispense viscous liquids at volumes from 20 -999 ml and has an anti-drip outlet that prevents food contaminating the sealing area of the container.

Dispensing filler for viscous liquids.

A semi-automatic piston filler can either have a row of filling heads, or they can be arranged in a circle as a 'carousel' filler. In operation, a piston draws product from a hopper into a cylinder and a rotary valve then changes position so that when the piston returns, the product is filled into containers. Piston fillers are relatively expensive to buy and are therefore suited to higher production volumes. This type of filler is less suitable for low-viscosity products, which can leak between the piston and cylinder.

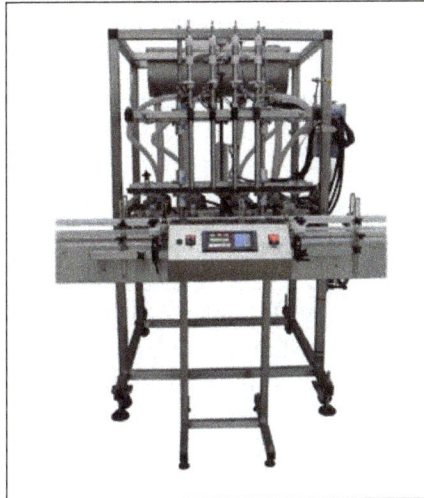

Piston filler.

Both low- and high-viscosity liquids may contain particles of food (e.g. fruit pulps, sauces and pickles) and these are more difficult to fill because simple liquid fillers described above may become blocked by particles. For manual filling, a wide-mouthed stainless steel or plastic funnel is more suitable, and products can be pushed through the funnel using a plastic plunger. Filling machines for high-viscosity liquids and liquids that have particles use servo pumps. Each filling nozzle has a dedicated servo-controlled positive displacement pump, but the capital cost of this equipment is high compared to other types of filler.

Hermetically sealed glass jars or metal cans that are used for bottled or canned foods are not filled completely. A 'headspace' is needed above the food to form a vacuum when it cools. When filling solid foods or pastes, it is necessary to prevent air from becoming trapped in the product, which would reduce the headspace vacuum. Viscous sauces or gravies are therefore filled before solid pieces of food. This is less important with dilute brines or syrups, as air is able to escape more easily before sealing.

## Sealing

Different types of closures for plastic and glass containers include metal or plastic caps and lids, and foil, plastic or paper covers. It is not possible in this Technical Brief to describe each of these closures in detail. In practice the selection of a package and closure depends mostly on their local availability and relative cost. The choice of packaging may therefore be a compromise between what is required and what is available/affordable, the penalty being a reduction in shelf life of the food.

## Pilferage and Tampering

Although total protection is not possible, tamper-resistant packaging delays entry into the package and tamper-evident packs indicate whether tampering has been attempted or has occurred. The main use of tamper resistance/evidence is for bottles, pots and jars that enable consumers to use the contents a little at a time and therefore need to be re-closable. Examples of tamper-evident or tamper-resistant closures are shown in table.

Table: Tamper-evident/-resistant packaging.

| Type of packaging | Tamper-evident or tamper-resistant features |
|---|---|
| Bottles and jars (glass or plastic) | Foil or membrane seals for wide-mouthed plastic pots and bottles |
| | Heat-shrinkable plastic sleeves for bottle necks, or bands or wrappers placed over lids. Perforated plastic or paper strips that must be cut or torn to gain access |
| | Breakable rings or bridges to join a bottle cap to a lower section (the container cannot be opened without breaking the bridge or removing the ring) |
| | Roll-on pilfer-proof (ROPP) caps for bottles (during rolling, a tamper-evident ring in the cap locks onto the bottle neck. A seal breaks on opening and the ring drops slightly) |
| | A safety button in lids for heat sterilised jars (a concave section formed in the lid by the headspace vacuum becomes convex when opened) |
| | A breakable plastic strip that shows if a jar has been opened |
| Flexible films | Film must be cut or torn to gain access |
| | Blister or bubble packs that show if the backing material has been separated from blisters. |
| | Laminated plastic/foil pouches must be cut to gain access |
| Tubes (aluminium Tubes or plastic) | Foil membrane over tube mouth that has to be punctured to gain access |
| Cans | Steel or aluminium cans are inherently tamper-resistant |

## Sealing Plastic or Glass Bottles and Jars

Bottle and jar closures can be grouped into three categories:

1. Pressure seals, used mostly for carbonated beverages. They include:

- Screw caps.

- Crimp-on lever-off ('crown' caps), or crimp-on screw-off steel caps. Crown caps can be sealed using the equipment.

- Roll-on screw-off aluminium caps. Roll-on screw caps can be fitted using a manual capper (also Roll-on-Pilfer-Proof (ROPP)) aluminium caps).

- Cork or polyethylene stoppers.

2. Normal seals, used for non-carbonated liquids (e.g. milk or wines):

- Pre-threaded, aluminium screw caps.

- Lug type screw twist-off steel caps. These 'Omnia' caps can be fitted using a simple manually operated capper.

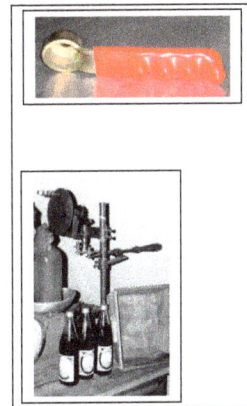

- Press-on, prise-off plastic caps.

- Push-in pull-out, or push-on pull-off caps, such as cork or synthetic stoppers. Corks can be inserted by hand after soaking overnight in clean water, or using a corking machine.

3. Vacuum seals, used for hermetically sealed jars:

- Screw-on twist-off or screw-on screw-off caps.

- Press-on prise-off, or press-on twist-off caps.

- Crimp-on prise-off caps.

Screw caps are usually hand fitted in small-scale operations.

In each type of closure the seal is formed by pressing a cushioning material against the rim of the container. The pressure must be evenly distributed to give a uniform seal around the rim. Typically, the cushioning material is made from plastic, cork or paperboard.

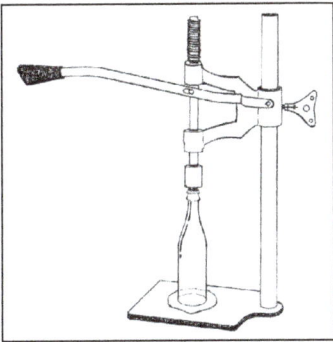

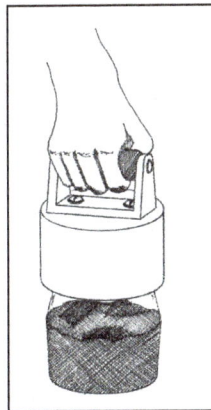

Manual roll-on screw capper.     Manual 'Omnia' capper.     Different types of Crown cappers.

Plastic bottles can have a variety of closures that have a pouring spout to dispense the contents. For example, caps can have a hinged top that reveals a dispensing opening in the cap. They are used for sqeezeable bottles (e.g. for creams, oils, sauces, mustard, mayonnaise or syrups).

Corking machine.

## Pot Sealers

A semi-automatic heat sealer is available for sealing film lids onto plastic pots at up to 100 pots an hour, but it is expensive and requires a source of compressed air. A cheaper and simpler sealer is available which will seal about 60 pots/hour. Alternatively an electric iron can be fitted to a suitable stand (e.g. a drill stand) and pressed down onto the surface of the pots to seal the film. Plastic trays are sealed with a plastic film or foil laminate that is heat sealed to the top flanges of the tray. Narrow-necked ceramic pots may be sealed with a cork stopper covered with candle-wax or beeswax.

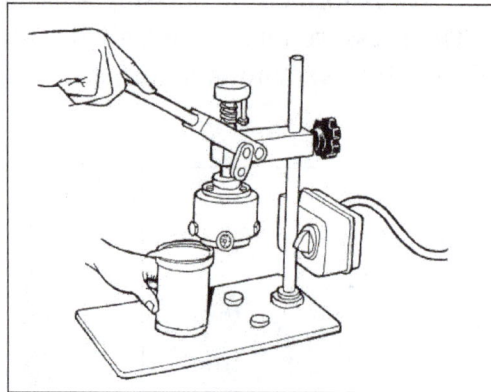

Pot sealer.

Wide-mouthed plastic pots and tubs or glass jars can be sealed by a range of different closures, including push-on, snap-on or clip-on lids (e.g. tubs for margarine or ice cream), and push-on or crimp-on metal or plastic caps (e.g. for nuts and snackfoods). These closures are not tamper-resistant, but containers can be fitted with tamper-evident aluminium foil or plastic membranes (e.g. yoghurt pots). Where a product is to be used over a period of time, or where additional protection is required for the membrane, a plastic clip-on lid may also be fitted to the pot. Rigid plastic or cardboard tubs for dry products can have a lid that opens so that the contents can be poured or shaken from the pack (e.g. small cardboard tubs for salt or spices). These 'disc top' closures have a plastic disc that is flipped up to reveal an opening. Special equipment is needed to seal metal cans.

## Sealing Pouches, Bags and Sacks Sealing Pouches, Bags and Sacks

### Plastic Films

Most plastic films are heat sealed but cold seals (adhesive seals) are sometimes used for heatsensitive products (e.g. chocolate, chocolate-coated biscuits or ice cream). To seal flexible films, the two surfaces of the film are heated until they partly melt and pressure is used to fuse the films together. The strength of the seal depends on the temperature, pressure and time of sealing. Although it is possible to seal plastic bags by folding the film over a used hacksaw blade and fusing it with a flame, the quality of the seal is variable and less attractive than using an electric heat sealer. A better (and faster) option is to use an electric bar-type heat sealer. If electricity is not available, it is possible to modify the sealer to heat the bar directly with a flame. A broader seal is formed with this equipment, which has better barrier properties and appearance. Sachets can be made by either buying film in the form of a tube, or by cutting the film and sealing the long side to make a tube. One end is sealed and foods are filled, before sealing the second end. It is important that foods do not stick to the inside of the pack where the seal is to made because they prevent a proper seal from forming or reduce its strength.

Hot-wire sealers have a metal wire that is heated to red heat to simultaneously form a seal and cut the film, whereas a bar sealer holds the two films in place between heated jaws until the seal is formed. In the impulse sealer, films are clamped between two cold jaws. The jaws are then heated to fuse the films and they remain in place until the seal cools and sets.

Bar sealer for plastic film.

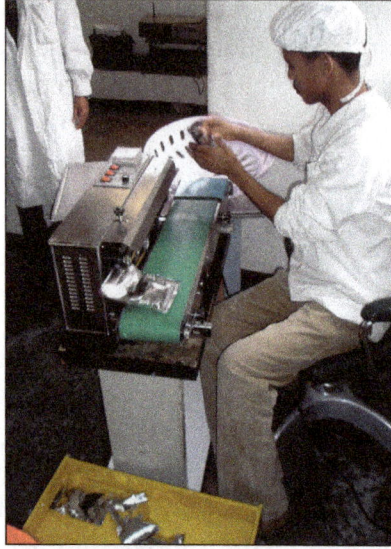

Band sealer for plastic film.

Rotary (or band) sealers are used for higher filling speeds. They have continuous belts that pass the mouth of a sachet or bag between heated rollers, and the two sides of the film are welded together. The seal may then pass through cooling belts that clamp it until the seal sets.

## Form-fill-seal (FFS) Equipment

FFS equipment has different forms: vertical form-fill-seal, known as 'transwrap' or 'flow pack', and horizontal form-fill-seal known as 'pillow pack' or 'flow wrap'. The machines use a roll of film, which they form into a tube, seal one end, fill the food and then seal the other end in a continuous operation. All types of FFS equipment are very expensive, require a source of compressed air and skilled maintenance technicians, and are therefore not likely to be affordable by small-scale processors.

## Shrink-wrapping and Stretch Wrapping

Low-density polyethylene is a film that shrinks in two directions when it is heated by either hot air or a radiant heater. Shrink-wrapped bottles, jars etc. are replacing cardboard distribution boxes in many countries.

Shrink-wrapped bottles and jars.

In stretch-wrapping, polyethylene or polyvinyl chloride film is wrapped under tension around boxes on a pallet. The equipment is simple and low-cost. Shrink-wrapping can also be used to manually wrap individual pieces of food (e.g. cheese) or to form the lid of plastic trays.

Stretch-wrapping cartons.

## Sealing Boxes and Cartons

Cardboard cartons or boxes that contain packaged foods are usually sealed by either hot-melt glue, or by adhesive tape. Simple tape dispensers are available, which make carton sealing faster and more efficient.

Adhesive tape dispenser.

## Sacks

Sack stitcher.

Sacks can be sealed by hand-stitching, but more commonly in commercial operations they are sealed using an electric sack-stitcher.

## Check-weighing and Metal Detection

Scales are used to ensure that fill-weights meet legal requirements and to minimise product give-away. They are tared to take account of the weight of the package and samples of products are taken from the production line to check that the net weight (i.e. the weight of product in the pack) is correct. At larger production rates, automatic check-weighers are fitted to product conveyor belts, where they weigh each pack and automatically reject those that are over/under-weight. These are however, very expensive and unlikely to be used at a small scale. Contamination with metal fragments can occur during processing as a result of wear or damage to equipment, and metal detectors are therefore an important part of quality assurance in all food processing. The basic components of a metal detector are a detection head, a conveyor that moves the products under the detection head, and a reject system that removes all packs that contain metal fragments. However, metal detectors are expensive and are often not used by small-scale processors.

Check-weighing scales.

## Storage

Food storage is an important component of food preservation. Many reactions that may deteriorate the quality of a food product occur during storage. The nutrient content of foods may be adversely affected by improper storage. For example, a significant amount of vitamin C and thiamine may be lost from foods during storage. Other undesirable quality changes that may occur during storage include changes in colour, development of off-flavours, and loss of texture. A properly designed food storage system allows fresh or processed foods to be stored for extended duration while maintaining quality.

The most important storage parameter is temperature. Most foods benefit from storage at a constant, low temperature where the rates of most reactions decrease and quality losses are minimized. In addition, foods containing high concentrations of water must be stored in high-humidity environments in order to prevent the excessive loss of moisture.

Careful control of atmospheric gases, such as oxygen, carbon dioxide, and ethylene, is important in extending the storage life of many products. For example, in the United States and Canada the apple industry utilizes controlled-atmosphere storage facilities in order to preserve the quality of the fruit.

# Food Industry Standards and Regulations

Globalization of processed foods markets has come about through consumer preferences, technical change, and other factors detailed throughout this report. Companies have responded to consumer demands, and there has been a scramble for standards bodies to catch up with commerce and impose some rules on the marketplace.

Demand for strict food standards in the United States is intensifying. The GATT Uruguay Round Agreement reduces trade protections and tightens rules for technical standards and investment barriers. The diminution of other forms of protection will raise the profile of product and process standards for both the free trade advocates and for those who demand protection. WTO members have stronger prospects for resolving disputes arising over product and process standards with ratification of the WTO than they did in the GATT arena prior to implementation of the Uruguay Round Agreement.

The motivations for standards are not always transparent or legitimate. To provide a brief illustration before proceeding to a more thorough discussion, the Mexican government in November 1994 began to update its food safety standards. Among them was a requirement that fluid milk could not be offered for sale more than 48 hours after pasteurization. U.S. milk bottlers in California, Arizona, New Mexico, and Texas, who were already selling fluid milk in Mexico, considered the short shelf life regulations as protectionist because they knew that continuously cooled milk has at least a 10-day shelf life. Devaluation of the Mexican peso diminished the U.S. milk shippers' commercial opportunity in the short run. But, they objected under North American Free Trade Agreement (NAFTA) rules similar to WTO rules that require that regulations have a scientific basis that is no more stringent than required to meet legitimate objectives.

## Product and Process Standards and Regulations

In a general sense, product and process standards fall within the broad category of competition policy, other examples of which include tied trade and marketing restrictions. The World Trade Organization (WTO), recognizes technical regulations, which are mandatory measures enforceable by law, and technical standards, which are voluntary measures. Governments and nongovernmental organizations such as industry associations can be involved in the development of standards, depending on the institutional relationships in a particular country. Both technical standards and regulations specify that a product must have certain characteristics or that certain processes must be followed in the manufacture of a product in order to qualify for import and sale. A product or process may be covered by standards in labeling laws, packaging laws, standards of identity, certification and inspection rules, and food safety standards.

The WTO defines technical regulations as: "document which lays down product characteristics or their related processes and production methods, including the applicable administrative provisions, with which compliance is mandatory. It may also include or deal exclusively with terminology, symbols, packaging, marking or labelling requirements as they apply to a product, process or production method".

The WTO definition of a standard is "document approved by a recognized body, that provides for common and repeated use, rules, guidelines or characteristics for products or related processes and production methods, with which compliance is not mandatory".

A typical example of product standards is a product definition. That is, products are required to be what they claim to be. For example, peanut butter has to be made from peanuts. Other examples include the Italian pasta purity laws and the Germany beer purity law, which strictly regulated the permissible ingredients in these products. In former times in Germany, beer could have only prescribed ingredients, and any other ingredients such as preservatives would make it illegal to import and sell such a product. The beer purity law is now mostly voided as far as trade is concerned, but it stood as a product standard for more than 450 years.

Examples of process standards include a ban on goods made with prison labor or a law against the importation of dairy products made of milk produced from cows treated with recombinant bovine somatotropin (rbST). At the time of writing, U.S. dairy products are not banned from any foreign market on the basis of the use of rbST in the United States. If a ban existed, this would be an example of a process standard. Some environmental standards are process standards. That is, there is no objection to the environmental effect of the product itself. The objection is to the exporter's competitive advantage that the lower environmental standards create.

## Effects of Standards and Regulations on Processed Food Firms

In the broadest terms, firms begin with the objective to maximize profits. If a firm decides to enter a foreign market, one might imagine that the decision to trade or invest—whether through new production facilities, acquisition of affiliates in the target market, licensing, or joint venture—is the result of a calculation of the relative costs of product placement in the foreign market. The cost of product placement is a more robust concept than transportation cost because the former is a more encompassing term than transportation costs, transactions costs, or even delivery costs. It involves many other costs including the costs of compliance with technical regulations on products and processes in the target market.

There are many considerations arising from product and process standards that affect a firm's decision to export or invest in foreign production. For instance, a firm can use any slack domestic production capacity to serve a foreign market. Often a firm has more and better information about production and marketing in its home market than in a foreign market. In a foreign market there may be a question of transparency, e.g., product and process standards may not be readily available or may change frequently or without notice. Some firms are quite protective of proprietary technology and formulations that a foreign government may require to be disclosed for safety reasons in order to be certified as eligible for import. A country may specify a minimum share of local content, which encourages domestic processing and the use of domestic materials in production.

Difficulty in meeting product and process standards for imports can lead a firm to buy manufacturing facilities in the foreign market it wishes to enter rather than attempt to export into that market. In a 1994 survey of multinational firms' decisions on export versus foreign direct investment, economies of scale and delivery costs relative to the value of the product were cited most frequently as decisive factors, but respondents also mentioned the following considerations that are included under product and process standards: inspection, certification, and risk.

## Effects of Standards and Regulations on Farmers

Even though most of the burden of compliance costs falls on food manufacturers, farmers in some cases are affected. The production agriculture sector can be affected by product and process standards in at least two ways. First, the utilization of domestic agricultural products in processed foods is an important component of the demand for these products.

Second, farmers are concerned with the compliance costs of process standards because they increase production costs—for example, a ban on growth-promoting hormones in beef production. However, the existence of high standards may result in greater consumption of U.S. food products if the standards help to create a perception of higher quality in the final products. The latter effect may or may not fully offset the higher costs of complying with standards.

## Effects of Standards and Regulations on Consumers

Obviously prices, quality, and safety of food are all very important to consumers. As a recent example of consumers' interest in food regulations, the Nutrition Labeling and Education Act of 1990 (NLEA) became effective on May 8, 1993, for regulation of health claims and on May 8, 1994, for regulation of nutrition labeling and nutri-ent content claims. The strict rules imposed by the NLEA may impose a substantial burden on foreign firms selling into the U.S. market that they may not have to meet in any other market. Aside from nutrition labeling, consumers want adequate stan-dards to ensure food safety and quality, and, for a given quality, they want to pay as little as possible. Consumer interests in food safety and food prices are not always congruent.

## Rationale for Product and Process Standards

There are many motivations for the imposition of standards, leading to considerable ambiguity. The first motivation in this discussion can be thought of as overall nation-al interests, including sovereignty, welfare, and distribution. Sovereignty arose as a concern during the Congressional debates on the GATT Uruguay Round and North American Free Trade Agreement. Some groups believed that U.S. laws, including standards for foods, should be outside the influence of pressure from other nations. Distributional issues include the effect of a new technology on farm structure. Wel-fare concerns include consumer protection and information issues such as labeling and food safety. Within the food safety category, sanitary and phytosanitary stan-dards (SPS) dealing with processed foods include pesticide residues and microbial contamination.

A second motivation for product and process standards is food safety concerns arising with new technologies. This raises the question of the degree to which the technolo-gies change the essential character of the product. In the case of recombinant bovine somatotropin (rbST), studies by the U.S. Food and Drug Administration (FDA) have found that the milk from untreated cows is identical to milk from cows treated with rbST. As a result, FDA has sharply restricted the wording that dairy foods marketers can use in labeling the product with respect to the use of rbST. The label cannot claim or imply safety or nutritional advantages for the non-rbST product. In the case of hor-mone implants in U.S. beef animals, a longstanding trade dispute has existed between the United States and the European Union, which bans the use of hormone implants that have withstood rigorous tests of safety, quality, and efficacy in the United States. In this instance, the EU does not claim that the meat is unsafe for human consumption, but objects to the process by which the beef was produced. Food irradiation presents another example of a technology that has generated some controversy based mainly on process. If a product with exactly the same food safety characteristics as an irradiated food could be produced without being irradiated, there would be no objection to the product.

A third motivation for standards is to facilitate desirable commercial developments. For instance, standards of identity, also known as product definitions, are defined by the FDA primarily for food safety purposes. Standards also facilitate trade by product description, reduce transactions costs, and improve market efficiency. Other standards serve commercial concerns including protection of geographical designations and brand names.

Fourth and finally, product and process standards can be motivated by trade protectionism. As, the Uruguay Round strengthened rules against the use of product and process standards as instruments of trade protection in the Agreement on Sanitary and Phytosanitary Measures and in the Agreement on Technical Barriers to Trade. In brief, the Uruguay Round outcomes insisted that standards be based on scientific evidence and appropriate risk analysis, that standards be transparent to other members, that standards be harmonized through international institutions where possible, and that members' standards, even if different from each other, be considered equivalent if the exporting country can demonstrate that the importing country's legitimate objectives are achieved by the exporting country's standard.

Out of a desire to protect their citizens, countries may inadvertently create an unjustifiable trade barrier. The Delaney Clause, which mandates zero tolerance for residues of pesticides that contain known carcinogens, is an example of a standard that could be challenged for not being based on appropriate risk assessment or appropriate science. For some known carcinogens, technological precision has advanced to the point that scientists can detect substances at harmless levels, called the "no observable effect level" (NOEL).

## Institutions

Product and process standards are not all governed by a single global body of rules. There are GATT rules, GATT precedents (case law), the Codex Alimentarius Commission, industry standards, and national institutions and laws whose jurisdictions overlap and contradict each other. The common functions of standards institutions are establishment of the standards, harmonization of standards across national and other administrative jurisdictions, enforcement of standards, and arbitration of disputes when members disagree on the application of standards. Of all the institutions performing these functions, the WTO has the greatest scope.

The WTO, as was also true for GATT before it, does not permit the use of technical standards as trade barriers. But Article XX of the GATT allows for general exceptions to the principles of most-favored nation and national treatment:

> "No country should be prevented from taking measures necessary to ensure the quality of its exports, or for the protection of human, animal or plant life or health, of the environment, or for the prevention of deceptive practices, at the levels it considers appropriate, subject to the requirement that they are not applied in a manner which would constitute a means of arbitrary or unjustifiable discrimination between countries where the same conditions prevail or a disguised restriction on international trade, and are otherwise in accordance with the provisions of this Agreement.".

Members' obligations following the Uruguay Round are fairly extensive, although there are in many cases no fixed measures for compliance. Members are encouraged to:

- Use existing international standards unless there are unusual circumstances;

- Participate in formulating new standards where none currently exist;

- Publish intent to create standards (where no international standard exists) so other countries have an opportunity to consult and suggest amendments before the standard is applied;

- Give higher priority to performance of standards in producing acceptable products than to design or description;

- Accept other countries' standards that differ from their own as long as the objectives of their own standards are met;

- Give notification of the objective and rationale of new technical standards and allow consultation; and

- Give assistance to other members (particularly developing country members) that want to establish technical standards.

In addition, the Technical Barriers to Trade (TBT) agreement in GATT contains a code of good practices for so-called conformity assessment procedures, which include the following:

> "Any procedure used, directly or indirectly, to determine that relevant requirements in technical regulations or standards are fulfilled. Conformity assessment procedures include, inter alia, procedures for sampling, testing and inspection; evaluation, verification and assurance of conformity; registration, accreditation and approval as well as their combinations.".

## Development and Harmonization of Standards

## Codex Alimentarius and other International Organizations

The SPS agreement specifically names three international institutions that have jurisdiction for establishing international standards. National governments provide official representation in these organizations. The Codex Alimentarius Commission, headquartered in Rome, is the main body establishing sanitary and phytosanitary standards. It began in 1962 with joint sponsorship of the U.N. Food and Agriculture Organization (FAO) and the U.N. World Health Organization (WHO) to establish food safety standards. Other smaller institutions include the International Office of Epizootics, established in 1924 to handle animal disease standards, and the International Plant Protection Convention, established in 1953 for plant health standards.

The International Standards Organization (ISO) and Comité Européen de

Normalization (CEN) are standards bodies operating on a regional or nongovernmental basis that can provide a competitive advantage to companies within the region where the standard is being developed. This is not to say that countries and companies outside of the organizations are being denied a voice in setting standards. The WTO rules call for members who are setting standards to provide opportunities for other countries to be consulted and to be given adequate time to comment before new standards are adopted, and to comply when new standards come into effect.

There remains a potential for trade diversion whether intended or not. For regional standards bodies such as CEN, the argument against regional harmonization as opposed to multilateral harmonization (i.e., through the WTO) is similar to the argument against regional trade liberalization agreements. A regional grouping (e.g., NAFTA, EU, or ASEAN) provides for common standards or mutual recognition within the trading group that may be preferential for members within the group, which means that an advantage is conferred to members and a disadvantage to nonmembers. Any impairment of access for products coming from outside the regional grouping may or may not be intentional, but standards may be set by compromises among the regional group's membership that do not consider the interests of nonmembers. The result may be a standard that requires greater costs of compliance for nonmembers than for members, thereby affecting trade patterns. In principle there is no problem with WTO rules as long as the trade agreement meets certain criteria (primarily, this means that the agreement covers substantially all trade), and that the standards are not set or applied in a discriminatory or arbitrary way. After all, the prime objective of regional trade agreements is to stimulate economic growth by facilitating trade within the region.

## ISO 9000: A Voluntary System of Standards

The International Standards Organization is located in Geneva, Switzerland. ISO 9000 is a method of quality assurance by which companies become certified as following recognized best practices. Certification declares to the buyer that the manufacturer has met a high quality standard. In other words, ISO 9000 standards are voluntarily followed practices, not technical standards required by a government. For industry standards, ISO 9000 is a system of quality assurance that can be used for food products, but is applied as well to products of other industries.

Widespread adoption of ISO 9000 standards could be followed by governmental recognition and adoption as minimum standards within a country or region and legally (or effectively) become an import standard, which would then be nondiscriminatory. It could be challenged as being a higher standard than can scientifically be shown as necessary. Bredahl and Zaibet conclude that "the momentum clearly seems to be in the direction of ISO adoption and the emergence over time of certification as a necessary condition to do business in the EU food sector."

## Public Policy Issues

The globalization of the processed foods market raises many public policy issues related to product and process standards. The increase in trade and foreign direct investment in processed foods increases public policy interest in standards not only because the commercial base is larger, but also because consumers are more demanding of high-quality, safe products at the least cost. Globalization has increased competition, and the rules are clearer and more enforceable. The completion of the Uruguay Round creates a clearer set of obligations regarding standards and, with the establishment of the Dispute Settlement Body, a stronger procedure for determining whether WTO members' food product standards serve only to support legitimate objectives. The following three public policy topics show the way toward establishing the importance of product and process standards, how and where they are used properly and improperly, and their effect on quantities of processed foods traded, funds invested internationally in the processed foods industries, and ultimately on food prices.

## Technical Standards as Trade Barriers

With clearer rules and a stronger method of dispute resolution, policy needs to be guided by an understanding of the prevalence of food standards acting as a foil for protectionist interests. Accordingly, there is a need to identify countries, products, or firms for which product and process standards may have been used improperly in the past. The process of identifying this problem will reveal where researchers and government officials should focus efforts to achieve the greatest benefit in terms of removing improper product and process standards that impede food trade. By examining the pattern of trade complaints brought to GATT, one should be able to ascertain the importance of product and process standards in trade rule violations and whether their importance is increasing or decreasing in number and as a share of all trade complaints. A further step to assess the use of technical standards as trade barriers would be to pair the records of trade complaints with patterns of FDI and trade in processed foods to determine if there are patterns in countries, products, or firms that would establish the prevalence of problems with standards. Additional information from industry sources could be valuable in identifying commercial concerns about the application of standards that never became formal trade disputes because of the lower likelihood of satisfactory resolution of disputes under the rules in force before the Uruguay Round Agreement.

## Harmonization versus Mutual Recognition

Two principles that govern standards in the WTO Technical Barriers to Trade Agreement and the Sanitary and Phytosanitary Standards Agreement are harmonization and equivalence (also known as mutual recognition). These principles are not always congruent. In some cases harmonization, rather than equivalence, is the guiding principle. In others, the reverse is true.

The varied use of the two principles leads to the question of whether it is possible to identify factors—institutional, economic, or political—that lead to the choice of one or the other. For example, Hooker and Caswell suggest that, for food trade, one should expect mutual recognition for quality standards and harmonization for food safety standards. Perhaps the type of product is an important factor in determining the guiding principle used under WTO. Harmonization may yield the greatest benefit for bulk or intermediate products that do not require significant processing. These products are more likely to be commingled and benefit more from the facilitation of packaging and handling, thereby lowering production or transaction costs. In contrast, harmonization of standards may not realize these benefits for products that have been further processed. Harmonization may impinge on consumer sovereignty by narrowing the spectrum of products offered to the consumer.

## Effect of Standards on Trade and FDI

Interviews have suggested that companies looking to enter a foreign market seldom give much consideration to "policies," including product and process standards, in deciding whether to enter that market via exports or FDI. To discover the importance of product and process standards in the trade versus FDI decision, one empirical approach would be to select cases and attempt to compare the relative costs of product placement associated with various approaches to entering a foreign market. Companies are understandably reluctant to divulge proprietary information about current decisions and operations, but perhaps suitable cases could be identified that would yield insight without compromising the firm's on-going operations. This approach would assess the impact of product and process standards on the costs, including the evaluation of risk, of product placement into a foreign market and thereby their influence on the method of entry into a market.

## References

- Importance-of-Materials-Handling-in-Food-Industry-6704463: ezinearticles.com, Retrieved 26 March, 2019
- Food-packaging: newworldencyclopedia.org, Retrieved 26 May, 2019
- Environmentally-compatible Food Packaging, Chiellini, E., Woodhead Publishing, Cambridge. 2008
- Packaging, food-preservation: britannica.com, Retrieved 25 June, 2019
- Small-scale Food Processing: A Directory of Equipment and Methods by Sue Azam-Ali Practical Action Publishing, 2003

# Permissions

All chapters in this book are published with permission under the Creative Commons Attribution Share Alike License or equivalent. Every chapter published in this book has been scrutinized by our experts. Their significance has been extensively debated. The topics covered herein carry significant information for a comprehensive understanding. They may even be implemented as practical applications or may be referred to as a beginning point for further studies.

We would like to thank the editorial team for lending their expertise to make the book truly unique. They have played a crucial role in the development of this book. Without their invaluable contributions this book wouldn't have been possible. They have made vital efforts to compile up to date information on the varied aspects of this subject to make this book a valuable addition to the collection of many professionals and students.

This book was conceptualized with the vision of imparting up-to-date and integrated information in this field. To ensure the same, a matchless editorial board was set up. Every individual on the board went through rigorous rounds of assessment to prove their worth. After which they invested a large part of their time researching and compiling the most relevant data for our readers.

The editorial board has been involved in producing this book since its inception. They have spent rigorous hours researching and exploring the diverse topics which have resulted in the successful publishing of this book. They have passed on their knowledge of decades through this book. To expedite this challenging task, the publisher supported the team at every step. A small team of assistant editors was also appointed to further simplify the editing procedure and attain best results for the readers.

Apart from the editorial board, the designing team has also invested a significant amount of their time in understanding the subject and creating the most relevant covers. They scrutinized every image to scout for the most suitable representation of the subject and create an appropriate cover for the book.

The publishing team has been an ardent support to the editorial, designing and production team. Their endless efforts to recruit the best for this project, has resulted in the accomplishment of this book. They are a veteran in the field of academics and their pool of knowledge is as vast as their experience in printing. Their expertise and guidance has proved useful at every step. Their uncompromising quality standards have made this book an exceptional effort. Their encouragement from time to time has been an inspiration for everyone.

The publisher and the editorial board hope that this book will prove to be a valuable piece of knowledge for students, practitioners and scholars across the globe.

# Index

www.ingramcontent.com/pod-product-compliance
Lightning Source LLC
Chambersburg PA
CBHW061957190326
41458CB00009B/2895